Sustainable Air Pollution Management: Theory and Practice

Sustainable Air Pollution Management: Theory and Practice

Contributors

Hsiao-Lan Liu, Yu-Sheng Shen et al.

AURIS
Reference

www.aurisreference.com

Sustainable Air Pollution Management: Theory and Practice

Contributors: Hsiao-Lan Liu, Yu-Sheng Shen et al.

Published by Auris Reference Limited

www.aurisreference.com

United Kingdom

Sustainable Air Pollution Management: Theory and Practice

ISBN: 978-1-78154-832-5

British Library Cataloguing in Publication Data

A CIP record for this book is available from the British Library

Printed in the United Kingdom

Exclusively distributed by CBS Publishers & Distributors Pvt. Ltd.

Sales & Distribution Rights only for India, Pakistan, Bangladesh, Sri Lanka, Nepal and Bhutan.This book is not to be sold outside these territories.

Contents

List of Abbreviations

ABL	Atmospheric Boundary Layer
AI	Aggregation Index
AMD	Acid Mine Drainage
AWMSI	Area-Weighted Mean Shape Index
BAEC	Bangladesh Atomic Energy Commission
BRT	Bus Rapid Transport
CAS	Chinese Academy of Sciences
CCN	Cloud Condensation Nuclei
CFC	Chloro Fluoro Carbon
COPD	Chronic Obstructive Pulmonary Disease
CSIR	Council for Scientific and Industrial Research
DRC	Democratic Republic of Congo
EF	Enrichment Factor
FB	Fractional Bias
FVC	Forced Vital Capacity
GDP	Gross Domestic Product
HHDI	High Human Development Index
IAP	Indoor Air Pollution
ICPMS	Inductively Coupled Plasma Mass Spectrometer
IPCC	Intergovernmental Panel on Climate Change
LHDI	Low Human Development Index
LOD	Limit of Detection
LOQ	Limit of Quantification
LPI	Largest Patch Index
MEP	Ministry of Environmental Protection
MHDI	Medium Human Development Index
MLR	Multiple Linear Regression
MSI	Mean Shape Index
NAAQS	National Ambient Air Quality Standard
NGO	Non-Governmental Organization
NMSE	Normalized Mean Square Error
NP	Number of Patches
OAP	Outdoor Air Pollution
OECD	Organization for Economic Co-operation and Development
PCA	Principal Component Analysis
PCR	Principal Component Regression
PD	Patch Density
PLS	Partial Least Squares
PM	Particulate Matter
PPP	Public-Private-Partnership
PRD	Pearl River Delta
PVC	Poly Vinyl Chloride

ROS	Reactive Oxygen Species
SD	System Dynamics
TSP	Total Suspended Particulates
UAV	Unmanned Air Vehicle
UV	Ultra-Violet
VC	Vital Capacity
VOC	Volatile Organic Carbons
WHO	World Health Organization
WRF	Weather Research and Forecasting
YRD	Yangtze River Delta

List of Contributors

Hsiao-Lan Liu
Department of Land Economics, National Chengchi University, Taipei 11605, Taiwan

Yu-Sheng Shen
Department of Land Economics, National Chengchi University, Taipei 11605, Taiwan

A.M.O. Abdul Raheem
Department of Chemistry, University of Ilorin, Ilorin, Nigera

F.A. Adekola
Department of Chemistry, University of Ilorin, Ilorin, Nigera

Xiaopeng Guo
School of Economics and Management, North China Electric Power University, Hui Long Guan, Chang Ping District, Beijing 102206, China

Xiaodan Guo
School of Economics and Management, North China Electric Power University, Hui Long Guan, Chang Ping District, Beijing 102206, China

Jiahai Yuan
School of Economics and Management, North China Electric Power University, Hui Long Guan, Chang Ping District, Beijing 102206, China

David O. Omole
Department of Civil Engineering, Tshwane University of Technology, Private Bag X680, Pretoria 0001, South Africa
Department of Civil Engineering, Covenant University, P.M.B. 1023, Ota, Ogun State +234, Nigeria

Julius M. Ndambuki
Department of Civil Engineering, Tshwane University of Technology, Private Bag X680, Pretoria 0001, South Africa

Dongyong Zhang
College of Information and Management Science, Henan Agricultural University, 15 Longzi Lake Campus, Zhengzhou East New District, Zhengzhou,

Henan 450046, China
Center for International Earth Science Information Network, The Earth Institute, Columbia University, P.O. Box 1000 (61 Route 9W), Palisades, NY 10964, USA

Junjuan Liu
College of Information and Management Science, Henan Agricultural University, 15 Longzi Lake Campus, Zhengzhou East New District, Zhengzhou, Henan 450046, China

Bingjun Li
College of Information and Management Science, Henan Agricultural University, 15 Longzi Lake Campus, Zhengzhou East New District, Zhengzhou, Henan 450046, China

Nicole Mölders
Geophysical Institute, University of Alaska Fairbanks, Fairbanks, USA
Department of Atmospheric Sciences, College of Natural Science and Mathematics, University of Alaska Fairbanks, Fairbanks, USA

Mary K. Butwin
Geophysical Institute, University of Alaska Fairbanks, Fairbanks, USA
Department of Atmospheric Sciences, College of Natural Science and Mathematics, University of Alaska Fairbanks, Fairbanks, USA

James M. Madden
Geophysical Institute, University of Alaska Fairbanks, Fairbanks, USA
Department of Atmospheric Sciences, College of Natural Science and Mathematics, University of Alaska Fairbanks, Fairbanks, USA

Huy N. Q. Tran
Geophysical Institute, University of Alaska Fairbanks, Fairbanks, USA
Bingham Entrepreneurship & Energy Research Center, Utah State University, Vernal, USA

Kenneth Sassen
Geophysical Institute, University of Alaska Fairbanks, Fairbanks, USA
Department of Atmospheric Science, University of Utah, Salt Lake City, USA

Gerhard Kramm
Geophysical Institute, University of Alaska Fairbanks, Fairbanks, USA
Engineering Meteorology Consulting, Fairbanks, USA

Yibin Cheng
Institute for Environmental Health and Related Product Safety, Chinese Center for Disease Control and Prevention, Beijing, China

Jiaqi Kang
Institute for Environmental Health and Related Product Safety, Chinese Center for Disease Control and Prevention, Beijing, China

Fan Liu
Institute for Environmental Health and Related Product Safety, Chinese Center for Disease Control and Prevention, Beijing, China

Bryan A. Bassig
Yale School of Public Health, New Haven, USA

Brian Leaderer
Yale School of Public Health, New Haven, USA

Gongli He
Institute for Environmental Health and Related Product Safety, Chinese Center for Disease Control and Prevention, Beijing, China

Theodore R. Holford
Yale School of Public Health, New Haven, USA

Ning Tang
Institute for Environmental Health and Related Product Safety, Chinese Center for Disease Control and Prevention, Beijing, China

Jian Wang
Chinese Center for Disease Control and Prevention, Beijing, China

Jian He
Gansu Provincial Center for Disease Control and Prevention, Lanzhou, China

Yanchang Liu
Institute for Environmental Health and Related Product Safety, Chinese Center for Disease Control and Prevention, Beijing, China

Yingchun Liu
Institute for Environmental Health and Related Product Safety, Chinese Center for Disease Control and Prevention, Beijing, China

Jiang Liu
Institute for Environmental Health and Related Product Safety, Chinese Center for Disease Control and Prevention, Beijing, China

Xun Chen
Institute for Environmental Health and Related Product Safety, Chinese Center for Disease Control and Prevention, Beijing, China

Heng Gu
Institute for Environmental Health and Related Product Safety, Chinese Center for Disease Control and Prevention, Beijing, China

Xiao Ma
West China School of Public Health, Sichuan University, Chengdu, China

Tongzhang Zheng
Yale School of Public Health, New Haven, USA

Yinlong Jin
Institute for Environmental Health and Related Product Safety, Chinese Center for Disease Control and Prevention, Beijing, China

Shukri I. Al-Hassen
Department of Geography, University of Basra, Basra, Iraq

Abdul Wahab A. Sultan
Technical College, Southern Technical University, Basra, Iraq

Adnan A. Ateek
Technical College, Southern Technical University, Basra, Iraq

Hamid T. Al-Saad
Department of Environmental Chemistry, University of Basra, Basra, Iraq

Salah Mahdi
Department of Environmental Chemistry, University of Basra, Basra, Iraq

Abdulzahra A. Alhello
Department of Environmental Chemistry, University of Basra, Basra, Iraq

Mahmoud M. M. Abdel-Salam
Department of Environmental Sciences, Faculty of Science, Alexandria University, Alexandria, Egypt

Md. Faridul Islam
Department of Chemistry, University of Dhaka, Dhaka, Bangladesh

Syada Sanjida Majumder
Department of Chemistry, University of Dhaka, Dhaka, Bangladesh

Abdullah Al Mamun
Department of Chemistry, University of Dhaka, Dhaka, Bangladesh
Department of Chemistry, University of Louisville, Louisville, USA

Md. Badiuzzaman Khan
Department of Environment Sciences Informatics and Statistics, Cà Foscari University of Venice, Venice, Italy
Department of Environment Science, Bangladesh Agricultural University, Mymensingh, Bangladesh

Mohammad Arifur Rahman
Department of Chemistry, University of Dhaka, Dhaka, Bangladesh

Abdus Salam
Department of Chemistry, University of Dhaka, Dhaka, Bangladesh

A. A. Marrouf
Department of Mathematics and Theoretical Physics, Atomic Energy Authority, Cairo, Egypt

Khaled S. M. Essa
Department of Mathematics and Theoretical Physics, Atomic Energy Authority, Cairo, Egypt

Maha S. El-Otaify
Department of Mathematics and Theoretical Physics, Atomic Energy Authority, Cairo, Egypt

Adel S. Mohamed
Department of Mathematics, Faculty of Science, Zagazig University, Zagazig, Egypt

xiii

Galal Ismail
Department of Mathematics, Faculty of Science, Zagazig University, Zagazig, Egypt

Preface

The text *Sustainable Air Pollution Management: Theory and Practice* discusses the fundamental aspects of traditional air pollution topics as well as some more advanced topics such as trans-boundary movement of air pollutants, air transportation of radioactive material, biological air pollutants, etc.

In first chapter, we examine the rapid development of green space change in the Taipei Metropolitan Area, and the impact of green space change on air pollution and on the microclimate. The aim of second chapter is to analyze environmental data gathered on the daily monitoring of ambient ozone, oxides of nitrogen, and sulfur (IV) oxide at five monitoring sites in Lagos and four monitoring sites in Ilorin, Nigeria. Third chapter analyzes the air pollution reduction policies' impact on the node enterprises of the thermal coal supply chain by comparing trends of key model factors with and without policy influence. Fourth chapter reviews developmental challenges confronting African countries with specific reference to the availability of potable water, sanitation, energy, water and ambient air. Fifth chapter aims to find the causes of the current severe air quality and explore the possible solutions by comparing China's air pollution regulations to that of the post London Killer Smog of 1952, in the United Kingdom (UK). Theoretical investigations on mapping mean distributions of particulate matter, inert, reactive, and secondary pollutants from wildfires by unmanned air vehicles (UAVs) have been performed in sixth chapter. The purpose of seventh chapter is to evaluate an indoor air pollution (IAP) intervention in rural areas in Gansu, one of the poorest provinces of China. Eighth chapter aims to analyze the geographic distribution of air pollutant concentrations in Basra Province, Southern Iraq, and to cartographically determine the spatial variation of air pollution levels as well as to recognize the hottest spots of air pollution. Ninth chapter focuses on the importance of personal exposure assessment to air pollution based on spatial and temporal activity patters both indoors and outdoors. Tenth chapter deals with trace metals concentrations at the atmosphere particulate matters in the Southeast Asian mega city. The influence of eddy diffusivity variation on the atmospheric diffusion equation has been discussed in last chapter.

Chapter 1

THE IMPACT OF GREEN SPACE CHANGES ON AIR POLLUTION AND MICROCLIMATES: A CASE STUDY OF THE TAIPEI METROPOLITAN AREA

Hsiao-Lan Liu and Yu-Sheng Shen

Department of Land Economics, National Chengchi University, Taipei 11605, Taiwan

ABSTRACT

In order to achieve a sustainable urban environment, the increase of green space areas is commonly used as a planning tool and adaptation strategy to combat environmental impacts resulting from global climate change and urbanization. Therefore, it is important to understand the change of green space areas and the derived impacts from the change. This research firstly applied space analysis and landscape ecology metrics to analyze the structure change of the pattern of green space area within the Taipei Metropolitan Area. Then, partial least squares were used to identify the consequences on microclimate and air pollution pattern caused by the changing pattern of green space areas within the districts of the Taipei Metropolitan Area. According to the analytical results, the green space area within Taipei Metropolitan Areas has decreased 1.19% from 1995 to 2007, but 93.19% of the green space areas have been kept for their original purposes. Next, from the landscape ecology metrics analysis, in suburban areas the linkages, pattern parameters, and space aggregation are all improving, and the fragmentation measure is also decreasing, but shape is becoming more complex. However, due to intensive land development in the city core, the pattern has becomes severely fragmented and decentralized causing the measures of the linkages and pattern parameters to decrease. The results from structural equation modeling indicate that the changing pattern of green space areas has great influences on air pollution and microclimate patterns. For instance, less air pollution, smaller rainfall patterns and cooler temperatures are associated with improvement in space aggregation, increasing the larger sized green space patch.

INTRODUCTION

The threat of disaster brought on by global climate change has captured the attention of most of the world's nations. Solutions to these threats fall primarily into two categories: mitigation and adaptation. The former emphasizes removing the causes of climate change to reduce the effects of the problems it poses. One result of this effect includes reducing the source of greenhouse gas emissions (or strengthening the sequestration of greenhouse gases), and thereby treating the root of the problem [1]. Current mitigation strategies target the reduction of greenhouse gas emissions in specific sectors (such as energy, industry, transportation, residential, commercial, *etc.*). The latter solution—adaptation—emphasizes responding and adjusting to the results of climate change, while reducing the damage it causes, possibly turning it to an advantage [2]. The current strategies of adaptation include comprehensively adjusting on a socio-economic level (such as strategies for land use, water resource management, public health and public construction, *etc.*). However, solutions of mitigation and adaptation compete with each other and conflict with regard to policy implementation and limited administrative resources [3,4]. These solutions can share benefits in a few special cases, such as in planting trees, developing and managing of green space, *etc.* In this paper, we will discuss the green space issues.

Green space, for the purposes of this paper, is open space covered by plants [5,6]. Green spaces are semi-natural areas [7] that not only have the environmental function of blocking noise [8], reducing carbon emissions and air pollution [9,10,11,12], conserving water and soil [13,14], adjusting the microclimate and moderating temperatures [12,15,16,17,18,19], but also have the ecological functions of recovering fertility, preserving ecologically sensitive areas, providing the habitat and feeding spaces for various species [20,21], and stabilizing ecological systems [22]. Moreover, green space has the landscape functions of buffering interferential land use while enhancing environmental beauty and visual aesthetics. It also has the socio-cultural functions of strengthening social cohesion and place identity by providing environmental education, recreation and cultural exchange [23,24,25]. Additionally, green space provides the health benefits of reducing tension and improving people's sense of satisfaction and happiness [26,27,28]. Thus, not only is green space the key to solving problems associated with climate change and over-urbanization, but it also plays a significant role in creating a sustainable urban environment that provides social and ecological balance. Therefore, the United Nations Conference on Sustainable Development (UNCSD), Organization for Economic Co-operation and Development (OECD), UKSDI (UKSDI is UK government sustainable development framework indicator) and Towards

Sustainable Europe all use green space as an important indicator for evaluation sustainable development.

The development and planning of green space can affect strategies of mitigation and adaptation simultaneously. Furthermore, changes to green space impact biological habitat, biodiversity, hydrologic cycle [29], soil properties [30], , and carbon storage [22]. Additionally, the impact of green space change on air pollution and microclimate is most important. Previous studies of air pollution and microclimate in urban area emphasize anthropogenic factors (such as building intensity, transportation, industrial development) [31,32,3 3,34,35,36,37,38,39,40]. There are few research papers that discuss the green space effects on air pollution and microclimates. Thus, the relationship among green space change, air pollution and microclimate is critical and calls for intensive study.

Previous research into the subject of green space has been exceptionally fruitful [9,10,11,22,29], but most studies have only evaluated the total areas of green space. However, when making a green space plan, not only should the total area be calculated, but also what is included within the same area must be considered. Should we plan big city parks or create a higher number of smaller community parks? How will their size and shape influence the functions of the green space? Past research has paid less attention to the structural change of green space and the impact of green space changes. Moreover, the spatial scale of analysis has been limited to the urban and community scale, seldom analyzing the districts within the large-scale metropolitan areas. Most natural development of green space is cross-border, thus analysis should include metropolitan districts and larger-scale geographic areas to acquire more reliable and valid results.

Because green space in metropolitan areas provides many functions, the development and planning of green space is an important way to mitigate and to adapt the environmental impact of climate change and over-urbanization while helping to achieve sustainable development goals. The level of urban development is different in each metropolitan area, and so the degree to which green space is impacted by development is different. In order to make green space planning effective, one must take into account the trends and impact of green space alterations. In addition, factors of air quality and microclimates are key for the residents' health. Because of this fact, green space alterations' impact on air pollution and microclimate requires closer examination. Therefore, for this paper, we will first apply spatial analysis and landscape ecology metrics to analyze structural changes in the pattern of green space within the districts of the Taipei metropolitan area. Then, we will use partial least squares to identify the impact on microclimate and air pollution patterns brought on by the changes to green space patterns.

This paper consists of six parts. The research motives and purpose, contents and previous research outcomes have been described in this section. The second part contains an outline of the research design, including the analytical framework, method and definition of variables, and hypotheses. The description of the empirical sample is provided in the third section. The analysis of the trend of green space changes is provided in the fourth part. The fifth section contains an analysis of green space changes' on air pollution and microclimates, and an explanation of the results of model calibrations as well as empirical analysis. The conclusions and suggestions are proposed in the final section.

RESEARCH DESIGN

Analytical Framework

The contents of various steps in this framework include the following (See also Figure 1):

- Defining the content and the spatial scope of green space changes: Confirm the research category and scope of green space change.
- Developing the research design: Determine the method, variables and hypotheses based on the theme and purposes of this study.
- Collecting and transferring the sample data: Collect and convert corresponding secondary data and cartographic material of the National Land-Use Survey from 1995 and 2007 through spatial analysis.
- Analyze the trend of green space change: Identify the state of green space change and migration between different land uses through land transfer matrix and spatial analysis. Additionally, analyze the spatial structure/composition change of green space through landscape ecology metrics.
- Discovering the impact of green space change on air pollution and the microclimate: Analyze the effect of green space change on air pollution and the microclimate, and identify the critical effect through partial least squares (PLS).
- Proposing conclusions and corresponding suggestions.

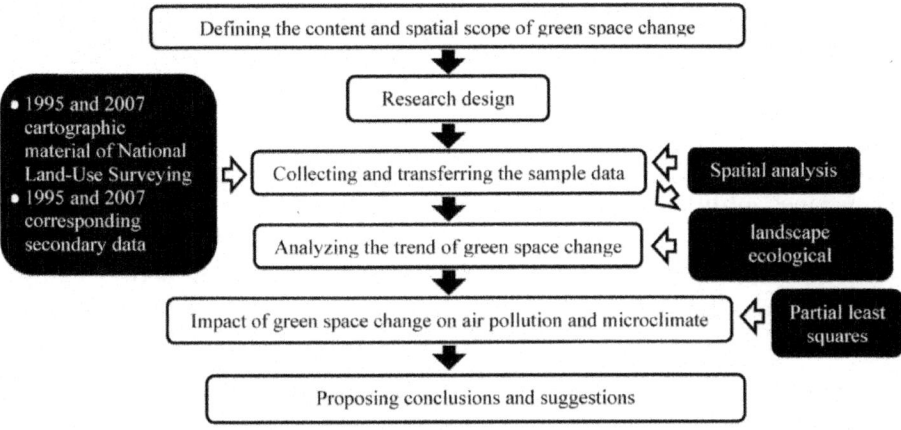

Figure 1. Analytical framework.

Method

Spatial Analysis

Spatial analysis is the method we employed to analyze spatial location, spatial distribution, spatial form, spatial space, spatial relationships and spatial change of object/events by topological, geometric, and geographic properties. The content of spatial analysis includes spatial data conversion and production, map rendering, exploratory data analysis, spatial statistics and simulation analysis [41,42].

For this paper, we used spatial analysis from the Geographic Information System (GIS) to process spatial data conversion, map rendering and spatial change analysis. With regard to spatial data conversion, we converted the spatial data from vectors into grid form, and re-categorized the land-use patterns of the Taipei Metropolitan Area from cartographic material provided via the National Land-Use Survey for 1995 and for 2007. Additionally, we used the Universal Kriging method (Universal Kriging method is Kriging with a local trend. The local trend or drift is a continuous and slowly varying trend surface on top of which the variation to be interpolated is superimposed. The local trend is recomputed for each output pixel and the operation is therefore more similar to the Moving Surface operation than to the Trend Surface operation [43,44].) to interpolate the missing data on air pollution and the microclimate in the districts of the Taipei Metropolitan Area. Through map rendering, this study illustrates green space change. For spatial change analysis, we identified the trends of green-space changes and migration between different land uses within the Taipei Metropolitan Area through a land transfer matrix, which was made by calculating transfer grids of land use from 1995 to 2007.

Landscape Ecological Metrics

Landscape Ecological Metrics can measure the context and structure of landscapes in different scales (such as patch level; class level and landscape level). This method is an important type of analysis utilized in landscape ecology (Landscape ecology emphasizes the interaction between spatial pattern, ecological process and scale, and focuses on the structure, function, change and management of landscape.) [45].

This paper uses Landscape Ecological Metrics to analyze the spatial structure of green space in the Taipei Metropolitan Area and its districts in 1995 and 2007. Moreover, it analyses the trend of green space change in the Taipei Metropolitan Area and its districts from 1995 to 2007.

Partial Least Squares

Partial Least Squares (PLS) is a form of structural equation modeling, distinguished from the classical method by being component-based rather than covariance-based.

The PLS algorithm is employed in PLS path modeling, a method of modeling a causal network of latent variables. The PLS model includes an inner model (The inner model is the part of the model that describes the relationships between the latent variables.) and an outer model (The outer model is the part of the model that describes the relationships between the latent variables and its observed variables.) [46]. PLS has an advantage in dealing with a reflective and formative model at the same time, strong predictive power, is suitable for analysis of small sample sizes, allows for analysis with multiple dependent variables and multiple independent variables, and avoids multicollinearity and limit on the sample distribution, such as dealing with interference data and missing data [46].

The sample size of this study is relatively small (only 48 administrative districts of Taipei metropolitan area), and the analytical model is reflective model. In addition, there are multiple dependent variables, multiple independent variables, and a complex relationship among those variables in the model. Thus, PLS is suitable for analyzing the impact of green space changes on air pollution and microclimate.

Because the sample size for this paper was relatively small, we adopted the Bootstrap Resampling Method for drawing 10,000 samples, and we used these samples to estimate the parameters and to verify our hypothesis. The

Bootstrap Resampling Method is a nonparametric method that was proposed by Efron [47], and it adopts resampling. Thus, even if the sample size of the PLS model is too limited, the PLS model can be estimated accurately through resampling methods.

Definition of Variables and Hypotheses

The empirical analysis includes the trend of green space change and the impact of green space changes on air pollution and the microclimate. Thus, the above mentioned variables and hypotheses will be defined in the following section.

Green Space Change Analysis

Analysis of green space change is achieved through such indicators as landscape ecology metrics. Landscape ecology metrics includes three levels: patch level, class level and landscape level.

The patch level is the sum of the grids, and is measured by calculating the characteristics of each patch (such as shape index, edge contrast index, *etc.*). The class level is the sum of a group of the same category of patches, and is indicated by calculating the characteristics of all types of classes (such as class area, core area, percentage of landscape, *etc.*). The landscape level is the sum of all patches or classes in the region, and is indicated by measuring the characteristics of all kinds of classes (such as Shannon's diversity index, relative patch richness, *etc.*) [45].

When examining the green-space change in the Taipei Metropolitan Area from 1995 to 2007, we used landscape ecology metrics of the class level for analysis. Since the indicators of landscape ecology metrics are numerous and complex, an explanation of some indicators requires repetition. Therefore, 14 front indicators of landscape ecology metrics were selected and analyzed for the purpose of research, such as Percentage of Landscape (PLAND), Number of Patches (NP), Patch Density (PD), Mean Patch Area (AREA_MN), Area-weighted Mean Patch Area (Area_AM), Largest Patch Index (LPI), Mean Shape Index (MSI), Area-weighted Mean Shape Index (AWMSI), Mean Nearest Neighbor Distance (ENN_MN), Area-weighted Mean Nearest Neighbor Distance (ENN_AM), Percentage of Like Adjacencies (PLADJ), Splitting Index (SPLIT), Radius of Gyration (GYRATE_MN), Area-weighted Radius of Gyration (GYRATE_AM), Clumpiness Index (CLUMPY), Aggregation Index (AI). The formula, units and methodology employed in measuring these 14 indicators are explained inappendix, Table A1.

Impact of Green Space Change

Definition of Latent Variables and of Observed Variables.

The purpose of analyzing the impact of green space change is to learn whether green space changes will affect the microclimate changes and air pollution changes, and if so, the degree of that influence. Therefore, we have provided the definitions of the perspective, the latent variables and the observed variables in Table 1.

The perspective includes three parts: "change of green space," "change of air pollution" and "change of microclimate."

The "change of green space" perspective includes six latent variables: "change of landscape," "change of fragmentation," "change of aggregation," "change of area," "change of proximity" and "change of largest patch percentage." With the exception of the observed variable of "changed area of maintaining and switching to green space," the other observed variables for measuring each latent variable are the changed rates of the landscape ecology metrics index.

The "change of air pollution" includes one latent variable: "change of air pollution emission." The observed variables for measuring the latent variable are the changed rates of different air pollutants, such as sulfur dioxide, nitrogen oxide, airborne particulate, carbon dioxide, nitric oxide and nitrogen dioxide.

The "change of microclimate" includes two latent variables: "change of rainfall type" and "change of temperature." The observed variable for measuring the "change of temperature" latent variable is the " change of mean annual temperature," and the observed variables for measuring the "change of rainfall type" latent variable are the " change of mean annual rainfall," "change of light rainy days," " change of torrential rainy days," and "change of non-rainy days". According to Taiwan's Climate Change Science Report [48] and the rainfall classification of the Central Weather Bureau, a standard of 0.1 mm $\leq$ daily precipitation <1.0 mm is defined as "light rainy day," a standard of daily precipitation ≥ 50.0 mm is defined as a "torrential rainy day," and a standard of daily precipitation <0.1 mm is defined as a "non-rainy day".

Table 1. The variables of PLS model.

Perspective	Latent Variables	Latent Variables Code	Observed Variables	Observed Variables Code
change of green space	change of Landscape	CL	• change of PLAND • changed area of maintaining and switching to the green space	cPLAND cWAERA
	change of fragmentation	CF	• change of NP • change of PD • change of SPLIT	cNP cPD cSPLIT
	change of aggregation	CA	• change of PLADJ • change of CLUMPY • change of AI	cPLADJ cCLUMPY cAI
	change of area	CR	• change of AREA-MN • change of AREA-AM	cAREA-MN cAREA-AM
	change of proximity	CN	• change of ENN-MN • change of ENN-AM	cENN-MN cENN-AM
	change of largest patch percentage	CP	• change of LPI	cLPI
change of air pollution	change of air pollution emission	CAP	• change of SO_2 emission • change of NO_x emission • change of PM emission • change of CO_2 emission • change of NO emission • change of NO_2 emission	SO_2 NO_x PM CO_2 NO NO_2

change of microclimate	change of rainfall type	CRT	• change of mean annual rainfall • change of light rainy day • change of torrential rainy day • change of non-rainy day	Rain lrd brd nrd
	change of temperature	CST	• change of mean annual temperature	Temp

Set of Hypothetical Relationship

The PLS model constructed for this study includes an outer model and an inner model. In the hypothetical relationship of the outer model, with the exception of the relationships between the "changes of rainfall type" the latent variables and the "change of mean annual rainfall," the "change of light rainy days" observed variables were negative, and the other relationships of latent variables and observed variables were positive. The hypothetical relationships of the inner model include the impact of green space change on air pollution change and the microclimate change, and the impact of air pollution change on the microclimate change (Table 2).

Table 2. Hypothetical relationship of latent variables in PLS model.

Endogenous latent variables/exogenous latent variables	Change of Landscape	Change of fragmentation	Change of aggregation	Change of area	Change of proximity	Change of largest patch percentage	Change of air pollution emission
change of air pollution emission	–	+	–	–	+	–	none
change of rainfall type	–	+	–	–	+	–	+/–
change of temperature	–	+	–	–	+	–	+/–

The above mentioned hypothetical relationships are as follows:

• The impact of green space change on air pollution change and microclimate changes.

The number and area of green space changes negatively affect changes in air pollution emissions and the microclimate. Thus, the "change of landscape" latent variables of the "change of green space" perspective are assumed to be opposite to the "change of air pollution emission," the "change of rainfall type" and "change of temperature" latent variables.

A large green space is synergistically helpful in reducing air pollution, the temperature and changes of rainfall type. Therefore, reducing the size of the green area and the percentage of the largest patch negatively impacts air quality, temperature and rainfall type. Keeping with the above statement, for this paper, we assumed the "change of air pollution emission," "change of rainfall type" and "change of temperature" latent variables were affected by the "change of area" and "change of largest patch percentage" latent variables of the "change of green space" perspective.

The aggregate effect of green space is the same as the scale effect of large green space; it can reduce air pollution, the temperature, and the change of rainfall type. Thus, the "change of aggregation" latent variables of the "change of green space" perspective are assumed to be opposite to the "change of air pollution emission," "change of rainfall type" and "change of temperature" latent variables.

The greater the nearest neighboring distance of green space, the more dispersive the patches and the less their effect in reducing air pollution, the temperature and the change of rainfall type. Moreover, the fragmentation of green space also has the same effect as the proximity of green space. Thus, the "change of fragmentation" and "change of proximity" latent variables of the "change of green space" perspective are assumed to be comparable to the "change of air pollution emission," the "change of rainfall type" and the "change of temperature" latent variables.

- The impact of air pollution change on microclimate change

Airborne particulate and sulfate aerosol reduce the volume of solar radiation and temperature through solar short wave radiation scattering (Because solar radiation enters into the atmosphere in the form of the short wave radiation, the more airborne particulate there are, the more the short wave radiation is reflected directly back into space. Thus, the above situation reduces the solar radiation reaching the earth surface). Air pollutants form easily in clouds, and clouds can reflect sunlight. In addition, clouds can warm through absorbing thermal radiation, as well as cool by diverting thermal radiation. The effect depends on the height and type of clouds [49,50,51,52,53]. Thus, the "change of air pollution emission" latent variables are assumed to affect the "change of temperature" latent variables.

Air pollutants are the source of cloud condensation nuclei (CCN). When air pollutants increase, the formation of rain is more difficult due to the number of cloud droplets increasing while the size of cloud droplets becomes smaller. Therefore, the formation of rain requires a substantially greater number of cloud droplets. Such a situation results in a change in the level of total rainfall and the number of rainy days, decreasing the number of light rainy days while increasing the frequency of torrential rainy days [52,54]. Thus, the "change of air pollution emission" latent variables are assumed to affect the "change of rainfall type" latent variables.

DATA

Empirical Area

In this paper, we are examining the rapid development of green space change in the Taipei Metropolitan Area, and the impact of green space change on air pollution and on the microclimate. In surveying the development of each metropolitan area in Taiwan, we found it apparent that the Taipei Metropolitan Area has attracted the largest population and greatest number of industries, and these developments have taken place rapidly. Thus, the Taipei Metropolitan Area is suitable as an empirical research subject. In this paper, the spatial scale of empirical analysis includes the 48 administrative districts of Taipei City, New Taipei City and Keelung City.

Description of the Empirical Sample

The empirical analysis focuses on the trend of green space change and the impact of green space change on air pollution and on the microclimate. Thus, the above empirical sample is described as follows:

- Green space change analysis

For this paper, we collected the empirical sample from the National Land-Use Survey for 1995 and 2007, and took the land use (attributes) of each grid in the empirical area by reclassifying land use categories through GIS technology (Figure 2 andFigure 3). The empirical sample uses the land transfer matrix and landscape ecological metrics as the input data for calculations. The data type of the empirical sample is the nominal scale.

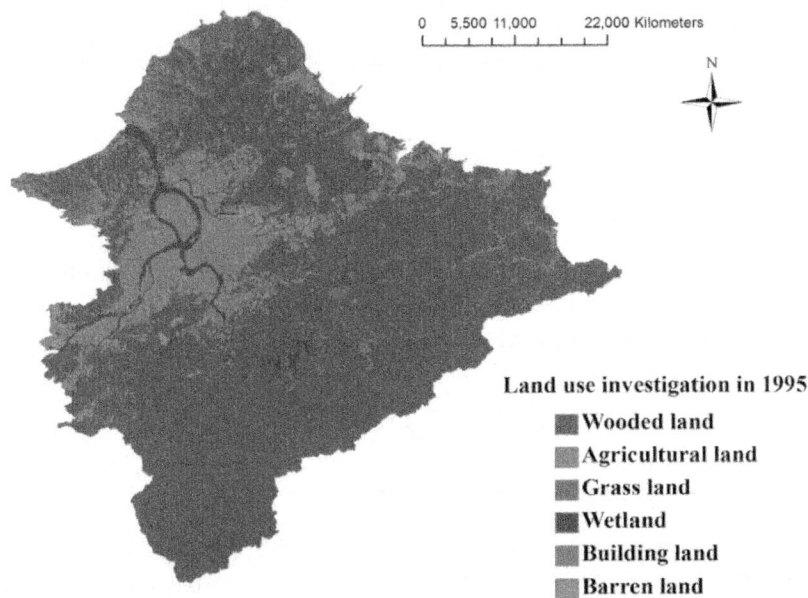

Figure 2. Taipei metropolitan area land use investigation in 1995.

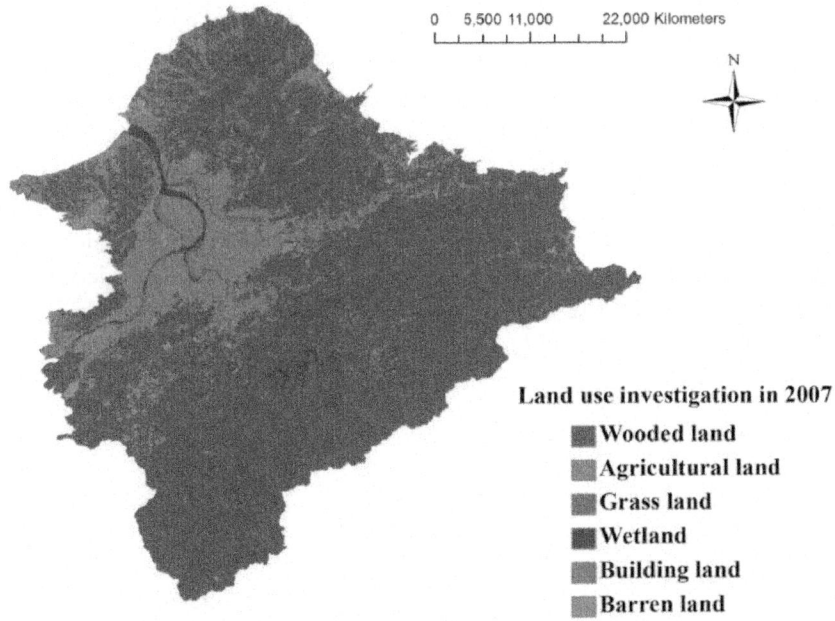

Figure 3. Taipei metropolitan area land use investigation in 2007.

• Impact of green space change.

In PLS modes, all empirical samples of observed variables are numerical data, and the units are percentages. The empirical sample of the "change of green space" perspective was collected from the National Land-Use Survey for 1995 and 2007, and was calculated by landscape ecological metrics and GIS technology. Additionally, the empirical sample of the "change of air pollution" and "change of microclimate" perspectives were gathered from the Environment Protection Agency for 1995 and 2007, and calculated by the Universal Kriging method to interpolate those missing data for air pollution and microclimates in the districts of the Taipei Metropolitan Area.

TREND OF GREEN SPACE CHANGE

States of Green Space Changes

According to the results of the green space transfer matrix in the Taipei Metropolitan Area from 1995 to 2007 (Table 3), green space area was reduced by 2339.5 hectares (occupies 1.19%), and 93.19% of the green space still retains its original use. According to the migration status, most green space is used for construction (occupies 4.87%), and increases in green space were created from wetland reclamation (occupies 32.38%) and barren land (occupies 38.15%), based on forestation and riverbank improvements.

Further analysis of green space in each sub-category of land use (such as wooded land, agricultural land, and grass land) finds that wooded land area has increased by 3377.5 hectares (occupies 2.03%), agricultural land area has decreased by 2323.5 hectares (occupies 12.05%) and grass land area has decreased by 3393.5 hectares (occupies 31.92%). According to the migration status of each sub-category of land use, 90.37% of the wooded land still maintains its original use, and the majority of agricultural land and grass land has been transfer to wooded land.

Table 3. Green space transfer matrix of Taipei Metropolitan Area from 1995 to 2007 (Unit: ha).

Land use area in 1995		Land use area in 2007							
		Green space				Wetland	Building land	Barren land	Total
		Wooded land	Agricul- tural land	Grass land	Total				
Green space	Wooded land	150,204.5	6524.5	2333.5	159,062.5	953.75	4989.25	1198	166,203.5
	Agricul- tural land	7917	6591.25	874	15,382.25	279	2975.25	648.5	19,285
	Grass land	5056.5	2036.25	1224	8316.75	186.5	1592.25	537	10,632.5
	Total	163178	15152	4431.5	182,761.5	1419.25	9556.75	2383.5	196,121
Wetland		1735.75	532.75	1326.5	3595	4877.5	1297.5	287.25	10,057.25
Building land		384.25	491.5	503.25	1379	283.75	25,376	1201.75	28,240.5
Barren land		1995.5	360.25	702.75	3058.5	188	2953.5	721.5	6921.5
Total		167,293.5	16,536.5	6964	190,794	6768.5	39,183.75	4594	241340.3
Area change		1090	−2748.5	−3668.5	−5327	−3288.75	10,943.25	−2327.5	-

Notes: 1. Left hand side represents the land use categories in 1995, and the land use categories in 2007 is on the top; 2. The number in the table represents the transferred areas from 1995 to 2007.

The districts of suburban and sub-core areas in the Taipei Metropolitan Area have decreased their maximum amount of green space, which change may have been caused by green space being transferred to public construction land or to large-scale residential land. In contrast, the built-up districts in the Taipei Metropolitan Area have preserved green space from development because the small amount of green space available is too limited to be advantageous to development, in addition to the fact that local residents guard it as a precious resource (Figure 4).

In summary, the green space area of the Taipei Metropolitan Area has been decreasing slightly, and most of the remaining green space is being transferred to the development of building projects. However, the area of green space has also increased from reclaimed wetlands (occupies 32.38%) and barren land (occupies 38.15%) based on forestation and riverbank improvements. Green space can still maintain its original function while not suffering serious damage from urban development.

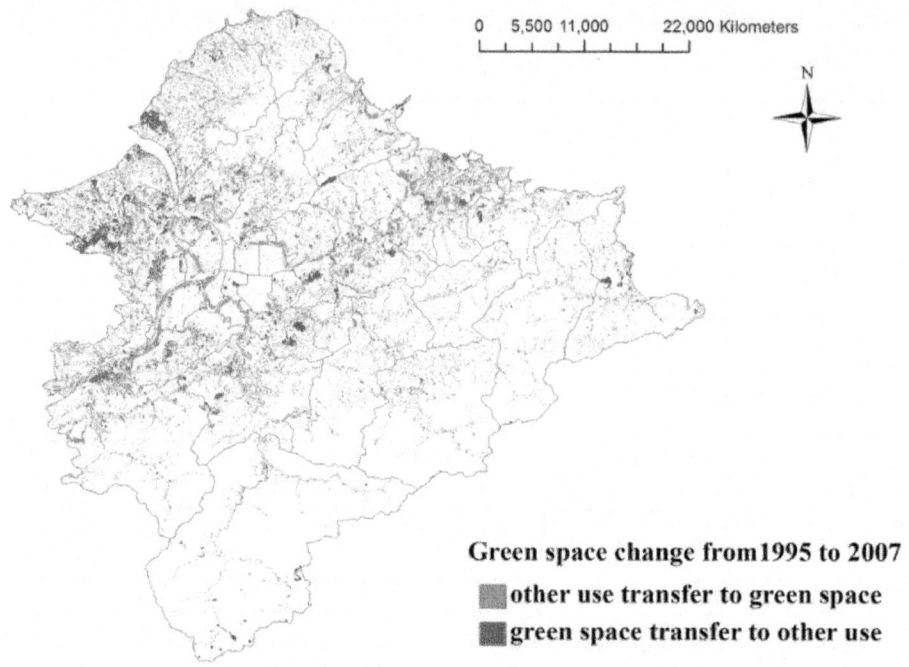

Figure 4. Taipei metropolitan area green space change from1995 to 2007.

Results of Landscape Ecological Metrics

In accordance with the results of landscape ecological metrics is shown in Table 4 and the spatial change status of green space in the Taipei Metropolitan Area from 1995 to 2007 is shown in Appendix.

As shown in Table 4 and Appendix (Figure A1), the total green space area has been decreasing slightly, the green space still has a large proportion (occupies 80%) of the overall landscape and plays a vital role. The number and density of the green space patches is incremental, pointing to the more fragmented trend of the green space. In the change of the green space patches area, it shows the area of the large patch increasing and the area of small patch decreasing. In the change of the green space shape, it shows the shape of the entire green space tending toward simple shapes, while the shape of the large patch tends toward a complex shape. In the change of the green space aggregation, it indicates green space is still highly centralized (above 90%). In the change of the green space proximity, it represents the distribution of the entire green space tending toward aggregation and the distribution of the large patch showing no change. In the change of the green space extendibility, it demonstrates the connection of the large patch increase and of the small patch reduction, as it closely relates to the area change of large and small patches.

According to the results of landscape ecological metrics, Table 4 and Appendix (Figure A2, Figure A3 and Figure A4) also show the sub-category changes of green space in the Taipei Metropolitan Area from 1995 to 2007.

Comparing landscape ecological metrics in change of the green space, the results of wooded land are similar except for the increase of the total wooded land area. It is not significant change of the wooded land proximity.

In light of the landscape ecological metrics for change of the agricultural land, the decrease of the total agricultural land area and its small proportion (occupies 7%) of overall landscape indicates that agricultural land is not in the primary position. The increasing of the number and density of the agricultural land patches, and the decreasing of the agricultural land patches area show a more fragmental trend with agricultural land from 1995 to 2007, and the large patch is more critical. In the change of the agricultural land shape, it indicates that the shape of all agricultural land is being simplified, closely related to an anthropogenic subdivision. In the change of the agricultural land aggregation, it means the centralization of agricultural land has been reduced, and it also closely relates to the anthropogenic subdivision. In the change of the agricultural land proximity, it represents the distribution of the entire agricultural land, and the large patch shows no change. In the change of the agricultural land extendibility, it means the connection between the entire agricultural land and the large patch is being reduced.

According to the landscape ecological metrics, the change of grass land, the decrease of the total grass land area and the small proportion of the overall landscape mean the grass land is not in the main position. Compared with landscape ecological metrics change of the agricultural land, the results for the grass land are similar. However, the grass land is more fragmented and decentralized than the agricultural land.

Table 4. Landscape ecological metrics results of green space in Taipei Metropolitan Area.

landscape ecological metrics	Green space		Wooded land		Agricultural land		Grass land	
	1995	2007	1995	2007	1995	2007	1995	2007
PLAND	80.11	79.15	67.89	69.27	7.88	6.93	4.34	2.96
NP	1433	2450	1531	2456	6313	9524	3918	5637
PD	0.59	1.00	0.63	1.00	2.58	3.89	1.60	2.30
AREA_MN	136.86	79.10	108.56	69.045	3.05	1.78	2.71	1.28
AREA_AM	104,529.1	164,396.1	99,271.33	148,504.8	149.67	112.62	111.06	79.44

LPI	55.81	72.85	51.28	64.8	0.4	0.38	0.22	0.21
MSI	1.31	1.26	1.36	1.25	1.30	1.23	1.24	1.15
AWMSI	19.59	36.59	30.11	51.33	4.24	4.06	3.30	2.73
PLADJ	95.41	93.90	93.25	91.80	62.26	52.52	64.31	51.58
ENN_MN	154.95	137.05	137.52	138.41	152.8	148.06	203.76	210.74
ENN_AM	100.65	100.35	100.63	100.41	119.2	120.48	148.80	162.10
SPLIT	2.92	1.88	3.63	2.38	20,767.1	31,379.9	50,762	104,239
GYRATE_MN	81.61	62.37	85.15	55.8	56.08	45.68	52.7	41
GYRATE_AM	14,120.05	18,448.3	14,408.84	18,302.39	472.2	414.56	413.53	429.23

IMPACT OF GREEN SPACE CHANGE

Model Calibrations and Verifications

For calibrating and verifying the impact model, we first analyzed the reliability and validity of the outer model, and then verified the explanatory power and significance of the path coefficient in the inner model. Finally, this paper verifies the goodness of the fit in the empirical model through goodness of the fit (GoF) index. (GoF index is a global criterion of goodness-of-fit, and was proposed by Tenenhaus *et al.* [55].)

• Outer model verification

In the test of reliability, Bollen [56] considered the significance of the coefficient value in which the t-value of observed variables must be greater than 1.96 ($\alpha = 0.05$, under two-tailed test) showed that the observed variable could reflect the meaning of the latent variable and that the outer model was suitable. The loadings of observed variables in this paper are significant, except for the "change of torrential rainy day" observed variables. This means that the most observed variable can reflect the meaning of the latent variable, and that the outer model is suitable (Table 5). The composite reliability (CR) in the empirical model is between 0.72 and 1 (Table 6); the value is higher than the standard (0.7) [57,58] proposes. The above results demonstrate the internal consistency of latent variables and meet the requirements of construct reliability.

In the test of validity; the square roots of the average variance extracted (AVE) for each latent variable are all greater than the correlation coefficient of the latent variable and other latent variables; which means the latent variable in the model has discriminating validity [59,60]. In addition; the composite reliability value of the latent variables is greater than 0.7; and the AVE value of latent variables are all greater than 0.5. As follows; the latent variables in the model have convergent validity [57,61].

Table 5. Results of loadings and test statistics in outer model development.

Perspective	Latent Variables	Observed Variables	Loadings	Z-value	Significance
change of green space	change of Land-scape	• change of PLAND	0.849	4.186	***
		• changed area of maintaining and switching to the green space	0.956	4.508	***
	change of fragmentation	• change of NP	0.933	9.461	***
		• change of PD	0.933	9.461	***
		• change of SPLIT	0.679	3.486	***
	change of aggregation	• change of PLADJ	0.996	11.101	***
		• change of CLUMPY	0.981	8.120	***
		• change of AI	0.996	11.110	***
	change of area	• change of AREA-MN	0.966	14.067	***
		• change of AREA-AM	0.952	14.170	***
	change of proximity	• change of ENN-MN	0.967	5.544	***
		• change of ENN-AM	0.963	4.508	***
	change of largest patch percentage	• change of LPI	1.000	-	
change of air pollution	change of air pollution emission	• change of SO_2 emission	0.888	30.560	***
		• change of NO_x emission	0.987	190.027	***
		• change of PM emission	0.920	34.084	***
		• change of CO_2 emission	0.873	53.975	***
		• change of NO emission	0.893	37.403	***
		• change of NO_2 emission	0.974	148.107	***
change of microclimate	change of rainfall type	• change of mean annual rainfall	-0.686	2.909	**
		• change of light rainy day	-0.895	4.211	***
		• change of torrential rainy day	0.296	1.819	
		• change of none rainy day	0.385	1.997	*
	change of temperature	• change of mean annual temperature	1.000	-	

Note: * $p < 0.05$, ** $p < 0.01$, *** $p < 0.001$.

Table 6. Goodness of fit in model development.

Perspective	Latent Variables	Cron-bach's α	CR [a]	Average	R Square
change of green space	change of Landscape	0.892	0.907	0.895	
	change of fragmenta-tion	0.818	0.854	0.841	
	change of aggregation	0.991	0.994	0.982	
	change of area	0.913	0.958	0.919	
	change of proximity	0.927	0.965	0.932	
	change of largest patch percentage	1.000	1.000	1.000	
change of air pol-lution	change of air pollution emission	0.965	0.972	0.883	0.479
change of micro-climate	change of rainfall type	0.701	0.718	0.699	0.328

Note: [a] CR means composite reliability.

- Inner model verification

According to the results of the inner model, the R^2 of "change of air pollution emission," "change of rainfall type" and "change of temperature" respectively are 0.4789, 0.3282 and 0.4656 (Table 6). According to the classification of explanatory power by Chin [62], the explanatory power for the impact of green space change on air pollution and on the microclimate is above the medium level.

Based on the significant test results of the path coefficient (Table 7), all latent variables in the "change of green space" perspective significantly affect the "change of air pollution emission" latent variable. The "change of landscape," "change of fragmentation," "change of area" and "change of air pollution emission" latent variables significantly affect the "change of rainfall type" latent variable. Finally, the "change of landscape," "change of fragmentation," "change of aggregation," "change of area," "change of largest patch percentage" and "change of air pollution emission" latent variables significantly affect the "change of temperature" latent variable.

Table 7. Empirical results of path coefficient in inner model.

Endogenous latent variables/exogenous latent variables	Change of Landscape	Change of fragmentation	Change of aggregation	Change of area	Change of proximity	Change of largest patch percentage	Change of air pollution emission
change of air pollution emission	−0.3619 **	0.2067 *	−0.2268 *	−0.2158 *	0.1161 *	−0.2211 *	-
change of rainfall type	−0.1162 *	0.1028 *	0.0088	−0.1042 *	0.0949	−0.1076	0.2726 *
change of temperature	−0.3523 *	0.2328 *	−0.1108 *	−0.1912 *	0.3148	−0.2022 *	−0.1707 *

Notes: $* p < 0.05$, $** p < 0.01$, $*** p < 0.001$.

- Goodness of fit in model

In this paper, the results have verified the goodness of fit in the empirical model through the GoF index as proposed by Tenenhaus *et al.* [55]. The GoF index was obtained as the geometric mean of the average communality index and the average R2 value. The GoF of the empirical model equals 0.61, a value higher than the 0.36 that Wetzels *et al.* [63] suggested, implying a suitable goodness of fit.

Results Analysis

The results of this study have verified the hypotheses and proposed the effects (such as direct effect, indirect effect, and total effect) of latent variables through PLS analysis (Figure 5). Among the effects, the indirect effect was calculated by the product-of-coefficients approach from Sobel [64], and tested the significance of the indirect effect with the Aroian test (The formula of Aroian test: (a×b)/SQRT(b2×S2a+a2×S2b+S2a×S2b), *a, b* is non-normalized coefficient, S2a, S2b is the standard errors of *a, b*.) [65,66]. The effects of the latent variables are stated respectively as follows (Figure 5):

- The impact of green space change on air pollution change and microclimate change

According to the empirical results of the "change of landscape" latent variables, the number and area of green space change negatively affects the change of air pollution emission and microclimate. The total effect of the green space area and the amount of air pollution change was −0.3619, and the total effect was derived entirely from the direct effect. The total effect of the green space area and the amount of rainfall patterns change was −0.1876, which summarizes the direct effect (−0.1162) and the indirect effect (−0.0714) from air pollution change. The direct effect of green space area and the number of temperature changes was −0.3523, and the total effect equaled the direct effect based on the statistical test of indirect effects not being significant.

According to the empirical results of the "change of area" and "change of largest patch percentage" latent variables, the changed size of the green area and the changed percentage of the largest patch negatively affect the air quality and the temperature. In addition, the changed size of the green area also negatively affects the rainfall type. Thus, the empirical results prove the existence of scale effect. The total effect of "change of area" and "change of largest patch percentage" on "change of air pollution emission" respectively were −0.2158 and −0.2211. The total effect of "change of area" and "change of largest patch percentage" on "change of temperature" respectively were −0.1912 and −0.2022. The total effect of "change of area" on "change of rainfall type" was −0.1042. Moreover, these influences were all entirely from the direct effect.

According to the empirical results of the "change of aggregation" latent variables, the aggregate effect of green space change negatively affects the change of air pollution emission and temperature. Thus, the empirical results prove the existence of the aggregate effect. The total effect of green space aggregation on air pollution change was −0.2268, and the total effect of green space aggregation on the temperature change was −0.1108. Furthermore, the influences were all entirely from the direct effect.

According to the empirical results of the "change of fragmentation" latent variables, the fragmentation of green space change positively affect the change of air pollution emissions and the microclimate. Thus, the empirical results are proof of the existence of the fragmented effect. The total effect of "change of fragmentation" on "change of air pollution emission," "change of rainfall type" and "change of temperature" respectively were 0.2067, 0.1028 and 0.2328. Additionally, the influences were entirely from the direct effect. According to the empirical results of the "change of proximity" latent variables, the connection of green space change positively affects the change of air pollution emissions. The total effect of "change of proximity" on "change of air pollution emission" was 0.1161, and the influences were all entirely from the direct effect.

In summary, the green space functions to reduce air pollutants, to reduce temperatures, and to improve rainfall types. Additionally, green space also has the effect of scale and aggregation on these functions. Thus, increasing the number, aggregation, area, scale/size of the green space and decreasing the fragmentation and proximity of green space can strengthen its function of reducing air pollutants, cooling temperatures and improving rainfall types.

- The impact of air pollution change on microclimate change

According to the empirical results, air pollution emissions changes negatively affect temperature change, which means the cooling effect is higher than the warming effect. This result may be the cause of solar short wave radiation scattering and cloud cooling. The total effect of "change of air pollution emission" on "change of temperature" was −0.1707, and the influences were entirely from the direct effect.

According to the empirical results, the air pollution emissions changes positively affect rainfall type change, with the total effect being 0.2726. This result means that higher emissions of air pollution would reduce the frequency of the light rainy days and the mean annual rainfall, while increasing the number of non-rainy days. This result may reflect the cause of increases in the number of cloud droplets and the scale of cloud droplets becoming smaller.

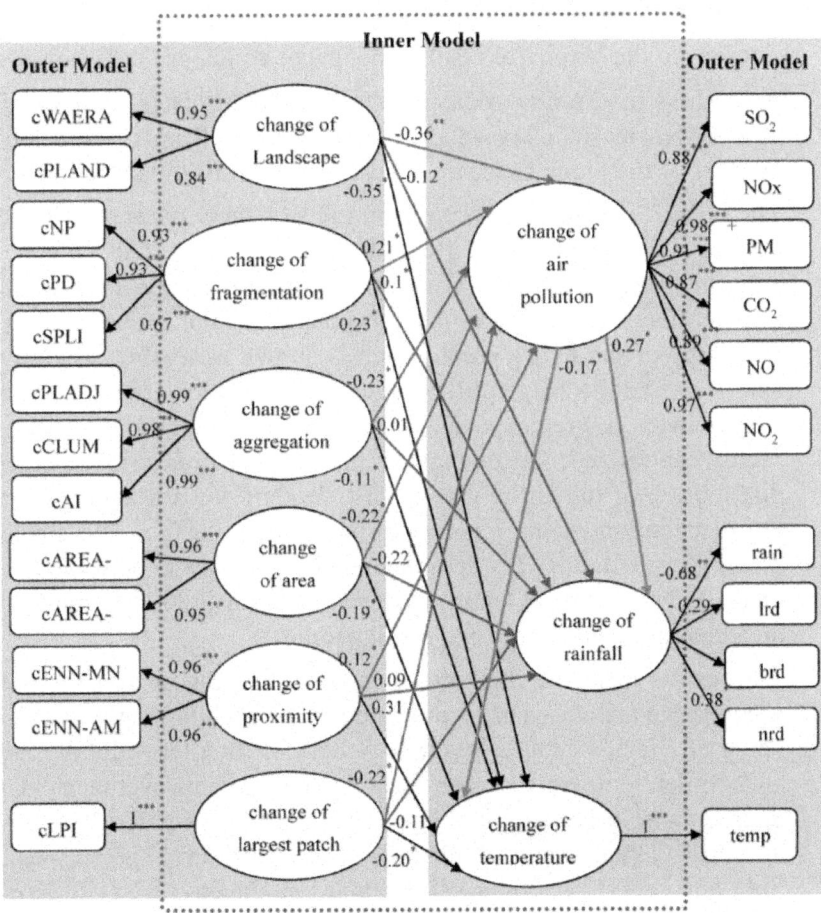

Figure 5. The calibration results of the modified model.

CONCLUSIONS AND SUGGESTIONS

Faced with the threats posed by global climate change while striving for the goal of sustainable development, increasing and conserving the green space area not only simultaneously achieves the effect both of mitigation and adaptation, but also helps in solving the problem of urbanization and non-sustainability. Recently, green space has often been transferred to other land uses because of the high level of urbanization. As a result, the environmental impact of reducing green space area is gradually becoming more significant. Therefore, for this research, we first applied space analysis and landscape ecology metrics to analyze the structural change of the pattern of green space areas within the 48 districts of the Taipei Metropolitan Area. We then used partial least squares to identify the consequences on microclimate and air pollution patterns caused by the changing patterns of green space area.

According to the empirical results, this paper concludes as follows:

- According to the analytical results, the green space area within the Taipei Metropolitan Areas has decreased 1.19% from 1995 to 2007, but 93.19% of the green space area have kept their original purposes.

- The results of landscape ecological metrics show the more fragmented trend of the green space. Regarding the change of the green space area, the large patch area is shown as increasing and the small patch area as decreasing. Regarding the change of the green space shape, the shape of the entire green space is shown as tending toward being simple, while the shape of the large patch tends toward being complex. The changes in green space aggregation mean that green space is still highly centralized. The changes in green space proximity represent the distribution of the entire green space tending toward aggregation, and the distribution of the large patch shows no change. The changes in green space extendibility demonstrate the connection between the large patch increase and the small patch reduction, as it closely relates to the area changes of the large and small patches.

- The results from the PLS model indicate that the changing pattern of green space area has a great influence on air pollution and microclimate patterns. For instance, less air pollution, smaller rainfall patterns, and cooler temperatures are associated with improvements in space aggregation, increasing the large sized green space patch. These results are similar to the research findings by Beatley [9], Jo [10], Yang *et al.* [11], Shin and Le [15], Herb *et al.*[16], and Leuzinger *et al.* [17]. However, they didn't discuss the patterns of green space, they only emphasized the total areas of green space. Although anthropogenic heat release is one of the important factors affecting air pollution and microclimates [32,36] in urban areas, we would assume that green space changes and anthropogenic heat release increase are major factors affecting the air pollution and microclimate pattern in cities.

- According to the results of the PLS model, the air pollution emissions change negatively affects the temperature change because of the cooling effect being higher than the warming effect. These results are similar to the findings by De Oliveira *et al.* [49], Koronakis *et al.* [50], and Wei and Hsu [52]. Moreover, the air pollution emission change positively affects the rainfall type change, which means the higher emission of air pollution reduces the occurrence of light rainy days and the mean annual rainfall and increases the number of non-rainy days. These results are consistent with Allen and Ingram [54].

Finally, in this paper we propose several suggestions based on the empirical results:

- The transfer of green space always occurs in the core area, in the sub-core area, and in the surrounding area. Thus, an urban growth boundary should be designated to avoid urban sprawl and the loss of green space.

- The trends and locations of wooded lands , agricultural lands and grass lands are different, and the conservation policies for green space should be adjusted for these different forms. The wooded land in suburbs should be effectively conserved to avoid decreases in the core areas and sub-core areas. Agricultural land should avoid arbitrary releases and transfers to land for development and construction while maintaining the integrity of the land. The grass land should be the focus of concern with regard to its change in the suburbs and sub-central area, and enhanced with conservation to avoid its loss.

- Green space has the function of reducing air pollutants, cooling temperatures and improving rainfall types. In addition, green space has the effect of scale and aggregation with regard to these functions. Thus, the development of the green space concept (such as a green city, garden city, green infrastructure) and policy (green space conservation) can be one of the solutions to climate change. Moreover, laws and governmental mechanisms relating to green space should be drafted immediately to help implement green policies.

- In this paper, we have only analyzed the changes from 1995 to 2007 due to limitations in data. If sufficient data can be provided in the future, these changes can be analyzed over a longer term.

ACKNOWLEDGMENTS

This research has been supported by NSC grant 101-2410-H-004-202, which is gratefully acknowledged.

AUTHOR CONTRIBUTIONS

Hsiao-Lan Liu is responsible for the conceptual design and actively involved in all steps of its implementation. Yu-Sheng Shen is responsible for developing the methodological framework.

APPENDIX

Table A1. The indicators of landscape ecology metrics (class level).

Indicator	Formula	Units	indicator content and measuring purpose
Percentage of Landscape (PLAND)	$\left(\sum_{j=1}^{n} a_{ij} \Big/ A\right)(100)$ a_{ij}: area of patch j (class i). A: total landscape area.	%	PLAND equals the percentage the landscape comprised of the corresponding patch type. The higher PLAND is, the more important the patch of corresponding class is. This paper use PLAND to analyze the importance of green space in landscape.
Number of Patches (NP)	n_i n_i: number of patches in the landscape of patch type (class) i.	None	NP equals the number of patches of the corresponding patch type. The higher NP is, the more fragmented the patch of corresponding class is. This paper use NP to analyze the fragmentation of green space in landscape.
Patch Density (PD)	$(n/A)(10{,}000)(100)$ n_i, A: definition as before.	Number per 100 hectares	PD equals the percentage the landscape comprised of the patches of the corresponding class. The higher PD is, the more fragmented the patch of corresponding class is. This paper use PD to analyze the fragmentation of green space in landscape.
Mean Patch Area (AREA_MN)	$\left(\sum_{j=1}^{n} a_{ij} \Big/ n_i\right)\left(\frac{1}{10000}\right)$ a_{ij}, n_i : definition as above.	Hectares	AREA_MN provides the measure of patch area in corresponding class. The higher AREA_MN is, the larger the patch of corresponding class is. This paper use AREA_MN to analyze the size of green space patch in landscape.

Area-Weighted Mean Patch Area (AREA_AM)	$\sum_{j=1}^{n}\left[a_{ij}\left(a_{ij}\Big/\sum_{j=1}^{n}a_{ij}\right)\right]\left(\dfrac{1}{10000}\right)$ a_{ij}: definition as before.	Hectares	AREA_AM provides the measure of area-weighted patch area in corresponding class. The higher AREA_AM is, the larger the patch of corresponding class is. This paper use AREA_MN to analyze the size of green space patch in landscape based on reducing the impact of small patches changes, and to compare with AREA_MN.
Largest Patch Index (LPI)	$(MAX(a_{ij})/A)(100)$ a_{ij}, A: definition as before.	%	LPI equals the percentage of the landscape comprised by the largest patch. The higher LPI is, the more important the patch of corresponding class is. This paper use LPI to analyze the contribution of largest patch, and to identify the advantage category in landscape.
Mean shape index (MSI)	$\sum_{j=1}^{n}\left(0.25\,p_{ij}\Big/\sqrt{a_{ij}}\right)\Big/n_i$ p_{ij}: perimeter of patch j (class i). a_{ij}, n_i: definition as above.	None	MSI provides the measure of patch shape in corresponding class. The higher MSI is, the more complex the patch shape in corresponding class is. This paper use MSI to analyze the complexity of the green space patch shape in corresponding class.
Area-Weighted Mean Shape Index (AWMSI)	$\sum_{j=1}^{n}\left[\left(0.25\,p_{ij}\Big/\sqrt{a_{ij}}\right)\left(a_{ij}\Big/\sum_{j=1}^{n}a_{ij}\right)\right]$ p_{ij}, a_{ij}: definition as above.	None	AWMSI provides the measure of area-weighted patch shape in corresponding class. The higher AWMSI is, the more complex the patch shape in corresponding class is. This paper use AWMSI to analyze the complexity of green space patch in corresponding class based on reducing the impact of small patches changes, and to compare with MSI.
Mean Nearest Neighbor Distance (ENN_MN)	$\sum_{j=1}^{n'}h_{ij}\Big/n'_i$ h_{ij} : distance between patch j (class i) to patch of the corresponding type; n'_i : number of patches in the landscape of patch type (class) i which having nearest neighbor distance.	Meter	ENN_MN provides the measure of patch distance in corresponding class. The higher ENN_MN is, the more dissipative the patch in corresponding class is. This paper use ENN_MN to analyze the proximity of each green space patches in corresponding class.

Area-Weighted Mean Nearest Neighbor Distance (ENN_AM)	$\sum_{j=1}^{n}\left[h_{ij}\left(a_{ij}\bigg/\sum_{j=1}^{n}a_{ij}\right)\right]$ h_{ij}, a_{ij} : definition as above.	Meter	ENN_AM provides the measure of area-weighted patch distance in corresponding class. The higher ENN_AM is, the more dissipative the patch in corresponding class is. This paper use ENN_AM to analyze the proximity of each green space patches in corresponding class based on reducing the impact of small patches changes, and to compare with ENN_MN.
Percentage of Like Adjacencies (PLADJ)	$\left(g_{ii}\bigg/\sum_{k=1}^{m}g_{ik}\right)(100)$ g_{ii}: number of like adjacencies between pixels of patch type (class) i based on the double-count method. g_{ik}: number of adjacencies between pixels of patch types (classes) i and k based on the double-count method.	%	PLADJ equals the percentage of cell adjacencies involving the corresponding patch type that are like adjacencies. PLADJ equals 0 when the corresponding patch type is maximally disaggregated and there are no like adjacencies. In contrast, The higher of PLADJ means high aggregation of the same type patch. This paper use PLADJ to analyze the aggregation of green space patches in corresponding class.
Splitting Index (SPLIT)	$A^{2}\bigg/\sum_{j=1}^{n}a_{ij}^{2}$ a_{ij}, A: definition as above.	None	SPLIT provides the measure separation of patch in corresponding class. SPLIT equals 1 when the landscape consists of single patch. SPLIT increases as the focal patch type is increasingly reduced in area and subdivided into smaller patches. This paper use PLADJ to analyze the separation of green space patches in corresponding class.
Radius of Gyration (GYRATE_MN)	$\sum_{r=1}^{z'}\left(h_{ijr}\big/z\right)$ h_{ijr}: distance between cell ijr (located within patch ij) and the centroid of patch ij (the average location), based on cell center to cell center distance. z: number of cells in patch ij.	Meter	GYRATE_MN provides the measure of patch connection and extendibility in corresponding class. The higher GYRATE_MN is, the more connected the patch in corresponding class is. This paper use GYRATE_MN to analyze the connection of green space patch in corresponding class.

Area-Weighted Radius of Gyration (GYRATE_AM)	$\sum\limits_{j=1}^{n}\left[\sum\limits_{r=1}^{z}\left(h_{ijr}/z\right)\left(a_{ij}/\sum\limits_{j=1}^{n}a_{ij}\right)\right]$ h_{ijr}, a_{ij}, z: definition as above.	Meter	GYRATE_AM provides the measure of area-weighted patch connection in corresponding class. The higher GYRATE_AM is, the more connected the patch in corresponding class is. This paper use GYRATE_AM to analyze the connection of green space patch in corresponding class based on reducing the impact of small patches changes, and to compare with GYRATE_ MN.
Clumpi-ness Index (CLUMPY)	$\begin{cases}(G_i - P_i)/(P_i) & for \quad G_i < P_i < 0.5 \\ G_i - P_i/1 - P_i & else\end{cases}$ $G_i = g_{ii}\Big/\left[\left(\sum\limits_{k=1}^{m}g_{ik}\right) - \min e_i\right]$ P_i: proportion of the landscape occupied by patch type (class) i. $\min e_i$: minimize the perimeter under the compact patch type (class) i of fixed grid number. g_{ii}, g_{ik}: definition as above.	None	CLUMPY provides the measure of patch aggregation in corresponding class. CLUMPY equals -1 when the focal patch type is maximally disaggregated. CLUMPY equals 0 when the focal patch type is distributed randomly, and approaches 1 when the patch type is maximally aggregated. This paper use CLUMPY to analyze the aggregation of green space patch in corresponding class.
Aggregation Index (AI)	$[g_{ss}/(\max \rightarrow g_{ss})](100)$ g_{ss}: number of like adjacencies between pixels of patch type (class) s based on the *single-count* method. $\max \rightarrow g_{ss}$: maximum number of like adjacencies between pixels of patch type (class) s based on the *single-count* method.	%	AI provides the measure of patch aggregation in corresponding class. AI equals 0 when the focal patch type is maximally disaggregated. AI increases as the focal patch type is increasingly aggregated and equals 100 when the patch type is maximally aggregated into a single, compact patch. This paper use AI to analyze the aggregation of green space patch in corresponding class.

Source: Leitão *et al.* [45], McGarigal and Mark [67].

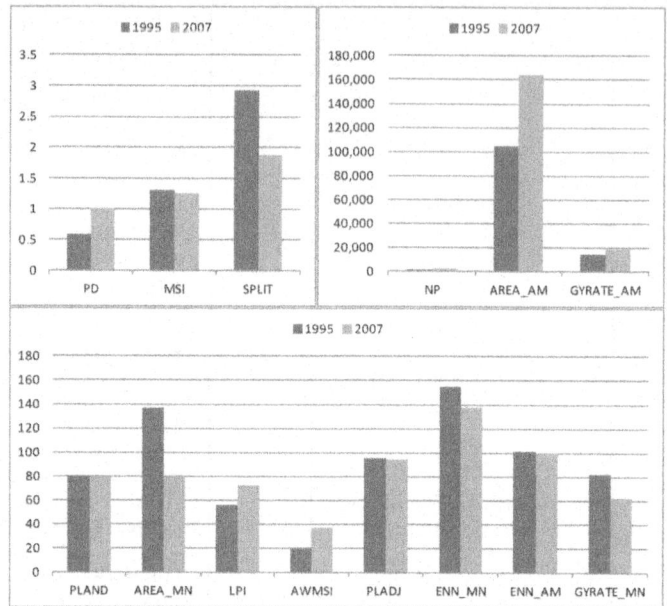

Figure A1. Change of landscape ecological metrics results from 1995 to 2007 (Green Space).

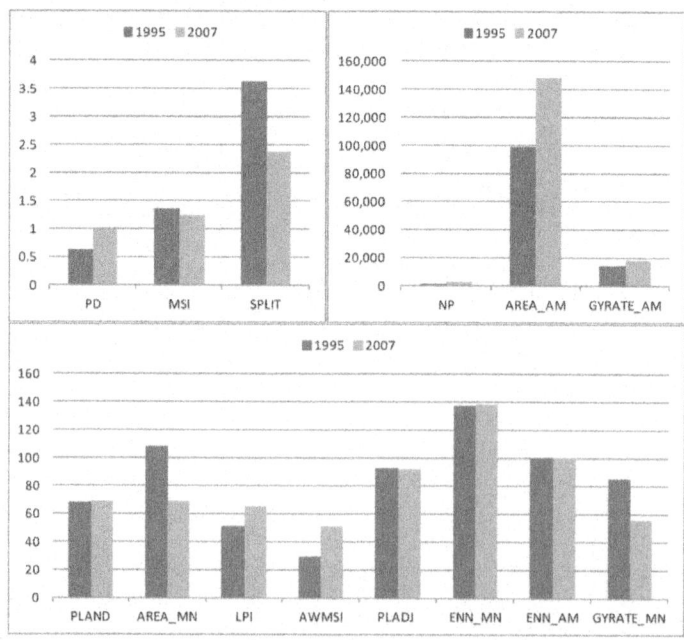

Figure A2. Change of landscape ecological metrics results from 1995 to 2007 (Wooded Land).

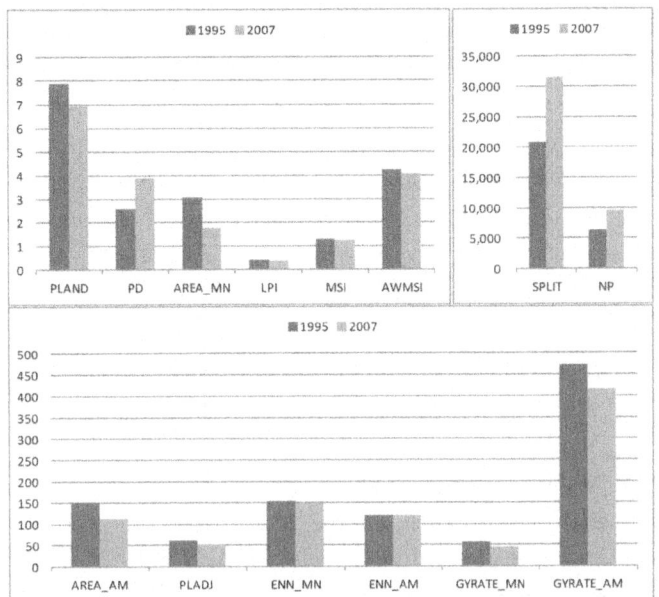

Figure A3. Change of landscape ecological metrics results from 1995 to 2007 (Agricultural Land).

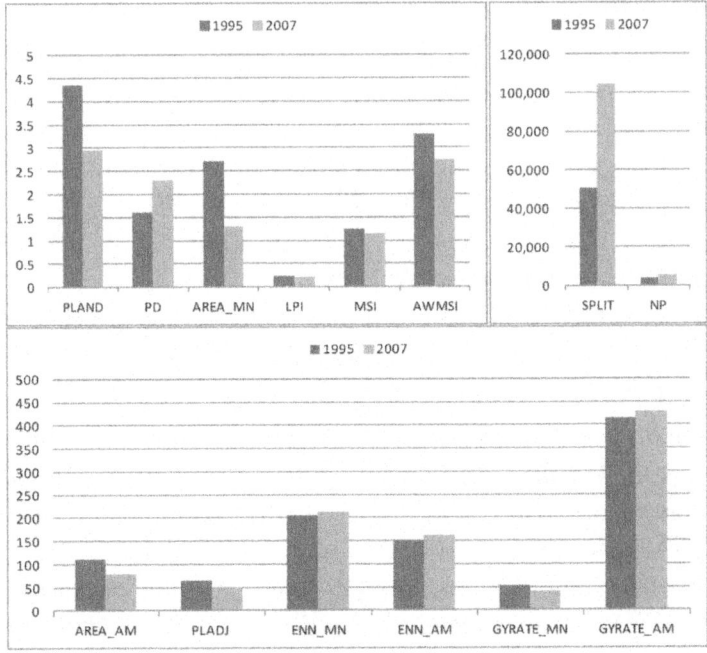

Figure A4. Change of landscape ecological metrics results from 1995 to 2007 (Grass Land).

REFERENCES

1. Parry, M.L.; Canziani, O.F.; Palutikof, J.P.; van der Linden, P.J.; Hanson, C.E. *Climate Change 2007: Impacts, Adaptation and Vulnerability—Working Group II Contribution to IPCC Fourth Assessment Report*; Cambridge University Press: Cambridge, UK, 2007.

2. Metz, B.; Davidson, O.R.; Bosch, P.R.; Dave, R. *Climate Change 2007: Mitigation of Climate Change—Working Group III Contribution to IPCC Fourth Assessment Report*; Cambridge University Press: Cambridge, UK, 2007.

3. Tol, R.S.J. Adaptation and mitigation: Trade-offs in substance and methods. *Environ. Sci. Policy* **2005**, *8*, 572–578.

4. Hunt, A.; Watkiss, P. Climate change impacts and adaptation in cities: A review of the literature. *Clim. Change* **2011**,*104*, 13–49.

5. Ahern, J. Planning for an extensive open space system: Linking landscape structure and function. *Landsc. Urban Plan.***1991**, *21*, 131–145.

6. Flores, A.; Pickett, S.T.A.; Zipperer, W.C.; Pouyat, R.V.; Pirani, R. Adopting a modern ecological view of the metropolitan landscape: The case of a greenspace system for the New York City region. *Landsc. Urban Plan.* **1998**, *39*, 295–308.

7. Linehan, J.; Gross, M.; Finn, J. Greenway planning: Developing a landscape ecological network approach. *Landsc. Urban Plan.* **1995**, *33*, 179–193.

8. Fang, C.F.; Ling, D.L. Investigation of the noise reduction provided by tree belts. *Landsc. Urban Plan.* **2003**, *63*, 187–195.

9. Beatley, T. *Green Urbanism: Learning from European Cities*; Island Press: Washington, DC, USA, 2000.

10. Jo, H.K. Impacts of urban green space on offsetting carbon emissions for middle Korea. *J. Environ. Manag.* **2002**, *64*, 115–126.

11. Yang, J.; McBride, J.; Zhou, J.; Sun, Z. The urban forest in Beijing and its role in air pollution reduction. *Urban For. Urban Green.* **2005**, *3*, 65–78.

12. Gill, S.E.; Handley, J.F.; Ennos, A.R.; Pauleit, S. Adapting cities for climate change: The role of the green infrastructure. *Built Environ.* **2007**, *33*, 115–133.

13. Pauleit, S.; Duhme, F. Assessing the environmental performance of land cover types for urban planning. *Landsc. Urban Plan.* **2000**, *52*, 1–20.

14. Miller, G.T.; Spoolman, S. *Environmental Science: Problems, Concepts, and Solutions*; Thomson Brooks/Cole: Pacific Grove, CA, USA, 2008.

15. Shin, D.H.; Lee, K.S. Use of remote sensing and geographical information systems to estimate green space surface-temperature change as a result of urban expansion. *Landsc. Ecol. Eng.* **2005**, *1*, 169–176.

16. Herb, W.R.; Janke, B.; Mohseni, O.; Stefan, H.G. Ground surface temperature simulation for different land covers. *J. Hydrol.* **2008**, *356*, 327–343.

17. Leuzinger, S.; Vogt, R.; Körner, C. Tree surface temperature in an urban environment. *Agric. For. Meteorol.* **2010**, *150*, 56–62.

18. Shashua-Bar, L.; Hoffman, M.E. The Green CTTC model for predicting the air temperature in small urban wooded sites. *Build. Environ.* **2002**, *37*, 1279–1288.

19. Shashua-Bar, L.; Hoffman, M.E. Vegetation as a climatic component in the design of an urban street: An empirical model for predicting the cooling effect of urban green areas with trees. *Energy Build.* **2000**, *31*, 221–235.

20. Song, I.J.; Hong, S.K.; Kim, H.O.; Byun, B.; Gin, Y. The pattern of landscape patches and invasion of naturalized plants in developed areas of urban Seoul. *Landsc. Urban Plan.* **2005**, *70*, 205–219.

21. Mathieu, R.; Freeman, C.; Aryal, J. Mapping private gardens in urban areas using object-oriented techniques and very high-resolution satellite imagery. *Landsc. Urban Plan.* **2007**, *81*, 179–192.

22. Whitford, V.; Ennos, A.R.; Handley, J.F. City form and natural process—Indicators for the ecological performance of urban areas and their application to Merseyside, UK. *Landsc. Urban Plan.* **2001**, *57*, 91–103.

23. Coley, R.L.; Kuo, F.E.; Sullivan, W.C. Where does community grow? The social context created by nature in urban public housing. *Environ. Behav.* **1997**, *29*, 468–494.

24. Thompson, C.W. Urban open space in the 21st century. *Landsc. Urban Plan.* **2002**, *60*, 59–72.

25. Chiesura, A. The role of urban parks for the sustainable city. *Landsc. Urban Plan.* **2004**, *68*, 129–138.

26. Grahn, P.; Stigsdotter, U.A. Landscape planning and stress. *Urban For. Urban Green.* **2003**, *2*, 1–18.

27. De Vries, S.; Verheij, R.A.; Groenewegen, P.P.; Spreeuwenberg, P. Natural environments-healthy environments? An exploratory analysis of the relationship between greenspace and health. *Environ. Plan.* **2003**, *35*, 1717–1731.

28. Gobster, P.H.; Westphal, L.M. The human dimensions of urban greenways:

Planning for recreation and related experiences. *Landsc. Urban Plan.* **2004**, *68*, 147–165.

29. Pauleit, S.; Ennos, R.; Golding, Y. Modeling the environmental impacts of urban land use and land cover change—A study in Merseyside, UK. *Landsc. Urban Plan.* **2005**, *71*, 295–310.

30. Meyer, W.B.; Turner, B.L. Human-population growth and global land-use cover change. *Annu. Rev. Ecol. Syst.* **1992**,*23*, 39–61.

31. Han, L.; Zhou, W.; Li, W.; Li, L. Impact of urbanization level on urban air quality: A case of fine particles (PM2.5) in Chinese cities. *Environ. Pollut.* **2014**, *194*, 163–170.

32. Civerolo, K.; Hogrefe, C.; Lynn, B.; Rosenthal, J.; Ku, J.Y.; Solecki, W.; Cox, J.; Small, C.; Rosenzweig, C.; Goldberg, R.;*et al.* Estimating the effects of increased urbanization on surface meteorology and ozone concentrations in the New York City metropolitan region. *Atmos. Environ.* **2007**, *41*, 1803–1818.

33. Li, B.; Wu, X. Economic structure and intensity influence air pollution model. *Energy Proced.* **2011**, *5*, 803–807.

34. Guttikunda, S.K.; Carmichael, G.R.; Calori, G.; Eck, C.; Woo, J.H. The contribution of megacities to regional sulfur pollution in Asia. *Atmos. Environ.* **2003**, *37*, 11–22.

35. Lindén, J.; Boman, J.; Holmer, B.; Thorsson, S.; Eliasson, I. Intra-urban air pollution in a rapidly growing Sahelian city.*Environ. Int.* **2012**, *40*, 51–62.

36. Arnfield, A.J. Two decades of urban climate research: A review of turbulence, exchanges of energy and water, and the urban heat island. *Int. J. Climatol.* **2003**, *23*, 1–26.

37. Zhong, S.; Yang, X.Q. Ensemble simulations of the urban effect on a summer rainfall event in the Great Beijing Metropolitan Area. *Atmos. Res.* **2015**, *153*, 318–334.

38. Kantzioura, A.; Kosmopoulos, P.; Zoras, S. Urban surface temperature and microclimate measurements in Thessaloniki. *Energy Build.* **2012**, *44*, 63–72.

39. Weng, Q. A remote sensing-GIS evaluation of urban expansion and its impact on surface temperature in the Zhujiang Delta, China. *Int. J. Rem. Sens.* **2001**, *22*, 1999–2014.

40. Landsberg, H.E. *The Urban Climate*; Academic Press: New York, NY, USA, 1981.

41. Fotheringham, S.; Rogerson, P. *Spatial Analysis and GIS*; Taylor & Francis: London, UK, 1994.

42. Fischer, M.M.; Getis, A. *Handbook of Applied Sapatial Analysis: Software Tools, Methods and Applications*; Springer-Verlag Berlin Heidelberg: Berlin, German, 2010.

43. Selby, B.; Kockelman, K.M. Spatial prediction of traffic levels in unmeasured locations: Applications of universal kriging and geographically weighted regression. *J. Transp. Geogr.* **2013**, *29*, 24–32.

44. Brus, D.J.; Heuvelink, G.B.M. Optimization of sample patterns for universal kriging of environmental variables.*Geoderma* **2007**, *138*, 86–95.

45. Leitão, A.B.; Miller, J.; Ahern, J.; McGarigal, K. *Measuring Landscapes: A Planner's Handbook*; Island Press: Washington, DC, USA, 2006.

46. Hair, J.F.; Hult, G.T.M.; Ringle, C.M.; Sarstedt, M. *A Primer on Partial Least Squares Structural Equation Modeling (PLS-SEM)*; Sage: Thousand Oaks, CA, USA, 2014.

47. Efron, B. Bootstrap methods: Another look at the jackknife. *Ann. Stat.* **1979**, *7*, 1–26.

48. National Science Council. *Climate Change in Taiwan: Scientific Report 2011*; National Science Council: Taipei, Taiwan, 2011.

49. De Oliveira, A.P.; Machado, A.J.; Escobedo, J.F.; Soares, J. Diurnal evolution of solar radiation at the surface in the city of São Paulo: Seasonal variation and modeling. *Theor. Appl. Climatol.* **2002**, *71*, 231–250.

50. Koronakis, P.S.; Sfantos, G.K.; Paliatsos, A.G.; Kaldellis, J.K.; Garofalakis, J.E.; Koronaki, I.P. Interrelations of UV-global/global/diffuse solar irradiance components and UV-global attenuation on air pollution episode days in Athens, Greece. *Atmos. Environ.* **2002**, *36*, 3173–3181.

51. Graedel, T.E.; Crutzen, P.J. *Atmosphere, Climate, and Change*; Scientific American Library: New York, NY, USA, 1994.

52. Wei, K.Y.; Hsu, H.H. *Global Change: An introduction*; Ministry of Education: Taipei City, Taiwan, 1997. (In Chinese).

53. Ding, Z.D. *Energy Crisis*; Wu-Nan Book Inc.: Taipei City, Taiwan, 2009. (In Chinese)

54. Allen, M.R.; Ingram, W.J. Constraints on future changes in climate and the hydrologic cycle. *Nature* **2002**, *419*, 224–232.

55. Tenenhaus, M.; Amato, S.; Esposito Vinzi, V. A global goodness-of-fit index for PLS structural equation modeling. Available online: http://www.

old.sis-statistica.org/files/pdf/atti/RSBa2004p739-742.pdf (accessed on 23 November 2014).

56. Bollen, K.A. *Structural Equations with Latent Variables*; Wiley: New York, NY, USA, 1989.

57. Fornell, C.; Larcker, D.F. Structural equation models with unobservable variables and measurement errors. *J. Mark. Res.* **1981**, *18*, 382–388.

58. Hulland, J. Use of partial least squares (PLS) in strategic management research: A review of four recent studies.*Strateg. Manag. J.* **1999**, *20*, 195–204.

59. Hair, J.F.; Anderson, R.E.; Tatham, R.L.; Babin, B.; Black, W.C. *Multivariate Data Analysis*; Prentice-Hall: Upper Saddle River, NJ, USA, 2006.

60. Esposito Vinzi, V.; Chin, W.W.; Henseler, J.; Wang, H. *Handbook of Partial Least Squares: Concepts, Methods and Applications*; Springer: Berlin, German, 2010.

61. Tabachnick, B.G.; Fidell, L.S. *Using Multivariate Statistics*; Allyn and Bacon: Boston, MA, USA, 2001.

62. Chin, W.W. The partial least squares approach for structural equation modeling. In *Modern Methods for Business Research*; Taylor & Francis: London, UK, 1998; pp. 295–336.

63. Wetzels, M.; Odekerken-Schröder, G.; van Oppen, C. Using PLS path modeling for assessing hierarchical construct models: Guidelines and empirical illustration. *MIS Q.* **2009**, *33*, 177–195.

64. Sobel, M.E. Some new results on indirect effects and their standard errors in covariance structure models. *Soc. Methodol.* **1986**, *16*, 159–186.

65. MacKinnon, D.P.; Krull, J.L.; Lockwood, C.M. Equivalence of the mediation, confounding, and suppression effect.*Prev. Sci.* **2000**, *1*, 173–181.

66. Shrout, P.E.; Bolger, N. Mediation in experimental and nonexperimental studies: New procedures and recommendations. *Psychol. Methods* **2002**, *7*, 422–445.

67. McGarigal, K.; Marks, B.J. *FRAGSTATS: Spatial Pattern Analysis Program for Quantifying Landscape Structure, USDA Forest Technique Report*; Pacific Northwest Research Station: Portland, OR, USA, 1995.

Chapter 2

AIR POLLUTION: A CASE STUDY OF ILORIN AND LAGOS OUTDOOR AIR

A.M.O. Abdul Raheem and F.A. Adekola

Department of Chemistry, University of Ilorin, Ilorin, Nigeria

INTRODUCTION

Air pollutants are continuously released from numerous sources into the atmosphere. Several studies have been carried out on the quantification of pollutants and analyzing their consequences on public health. It has been estimated that each year between 250 and 300 million tons of air pollutants enter the atmosphere above the United States of American [Dara, 2004; Onianwa, 2001; Stephen and Spencer, 1992]. Tropospheric pollution causes degradation of crops, forests, aquatic systems, structural materials, and human health. It was reported recently, that NO_x air pollution is becoming a far reaching threat to USA National Parks and Wilderness Areas as these areas are suffering from harmful effects of oxides of nitrogen pollution [EDFS, 2003]. It has also been confirmed that NO_x contributes to ground – level ozone (smog) pollution which can cause serious respiratory problems, especially young children and the elderly, as well as healthy adults that are active outdoors. Furthermore, the same report confirmed worsening ozone concentration in nearly all the national parks over the last ten years [EDFS, 2003]. Towards this, an assessment of new vehicles emission certification standards was carried out in metropolitan area of Mexico city and the results show that light duty gasoline vehicles account for most carbon (II) oxide and NO_x emissions [Schifter et al, 2006]. The European Environmental Agency also reported very recently that more than 95% contribution to nitrogen oxides emission to the air comes from fuel combustion processes from road transport, power plants and industrial boilers [EEA, 2006]. There is reported evidence of average chronic damage to the human lung from prolonged ozone exposure [EEA, 2006]. Sulphur in coal, oil and minerals are the main source of the

Sulphur (IV) oxide in the atmosphere. Moreover, peak concentrations above European Union limit still occur, especially close to point sources in the cities. Asian cities have some of the highest levels of air pollution in the world. In Asia, hundreds of thousands of people in urban areas get sick just by breathing the air that surrounds them. However, the WHO 2006 estimates that dirty air kills more than half a million people in Asia each year of which burden falls heaviest on the poor as reported by Ogawa, 2006. The worsening of the situation has been attributed to cumulative effects of rapid population growth, industrialization and increased use of vehicles. The ozone primary tropospheric pollutants (SO_2, NO_x, HC_s, and CO) often react in the atmosphere to form secondary pollutants which are acidic compounds (H_2CO_3, H_2SO_4 and HNO_3) and photochemical oxidants. Environmental damage frequently results from several primary and secondary pollutants acting in concert rather than from a single pollutant. Tropospheric oxidants such as ozone, PBN, PAN illustrate the complexity of atmospheric chemistry and processes. They help to form acidic compounds thereby contributing to green house warming and hence, damage to human health, animal, plant life and materials [USEPA, 1998; Dara, 2004]. Significant changes in stratospheric ozone, high above the troposphere, can affect tropospheric oxidants level [USEPA, 1998]. If increased UV-B radiation penetrates a depleted ozone shield, the photochemical formation of ground level oxidants may be enhanced [Stoker and Seager, 1972]. Green house warming could amplify this effect: A study carried out in three U.S. cities; Nashville, Philadelphia, and Los Angeles showed that a large depletion of stratospheric ozone, coupled with the green house warming, could increase smog formation by as much as 50% [Adelman, 1987]. The study also showed that NO_2 concentration might increase more than ten folds. There is however progress towards the reduction of anthropogenic emissions of NO_x, CO, volatile organic compounds in Europe and North American [Jonson et al, 2001]. However, the concentration of air pollutants emitted into the atmosphere is on the increase in the Southeast Asia and other parts of the World [Jonson et al, 2001]. It is therefore expected that the emissions from Africa and other parts of the Worlds that are yet to take strict and effective controlling measures on emissions will influence the free tropospheric levels in most of the Northern Hemisphere.

During a five-day period marked by temperature inversion and fog in London in 1952, between 3,500 and 4,000 deaths in excess of normal occurred with 1.3 ppm SO_2 level recorded [O'Neill, 1993; ACGIH, 1991]. SO_2 is oxidized to SO_3 in the atmospheric air by photolytic and catalytic processes involving ozone, oxides of nitrogen (NOx), and hydrocarbon (HC), giving rise to the formation of photochemical smog [Dara, 2004].

In contrast to NO_2, SO_2 is deleterious to plant life [Haagen-Smith, 1952; Molski and Dmuchowski, 1986]. Air pollution causes the decline in Eastern Europe [Nihlgard, 1985; Schutt and Cowling, 1985]. Area with extensive large-scale forest decline correlate with the areas where SO_2 concentration were elevated [ECE, 1984; Molski and Dmuchowski, 1986], therefore If the concentration of SO_2 is higher along with other gaseous pollutants (NO_x, O_3, HC, organic and inorganic peroxides etc) in the troposphere and continues to accumulate over time, the overall concentration can have a negative effect on health, vegetation and structures [Abdul Raheem, 2007; Abdul Raheem et al., 2009a].

Surface concentration of ozone has been reported to be on the increase in the last decades in northern hemisphere [Vautard et al., 2007] and in Southeast Asia [Jaffe, 1999]. A rise of about 26% in the ambient concentration of ozone has been reported in Taiwan between 1994 and 2003 [Chou et al., 2006]. High levels of ambient air ozone can cause serious damage to health. The health hazards include shortness of breath, nausea, eye and throat irritation, and lung damage [Menezes and Shively, 2001]. Identification of air pollution source characteristics is an important step in the development of regional air quality control strategies. Receptor modeling, using measurements of pollutant concentrations at one or more sample sites, is often a reliable way to provide information regarding source regions for air pollution [Watson, 1984]. One of such receptor-modeling technique is principal component analysis (PCA) [Einax and Geiss, 1997; Jackson, 1991; Norman, 1987]. This is often combined with multiple linear regression (MLR), principal component regression (PCR), and partial least square regression, which have been demonstrated as powerful tools for handling several environmental problems, especially source apportioning [Otto, 1999; Timm, 1985; Vogt, 1989].

In recent years, certain statistical techniques that incorporate the influence of meteorological variables have been applied to asses the trend in ozone levels in the ambient air [Bakken et al., 1997]. One common approach is the use of a parametric regression model to link some characteristics of ozone, such as the mean level of ozone to meteorological variables. Other scientists have equally used PCA to pattern the spatial and temporal variations of ozone and to identify the important factors influencing ozone concentration [Klaus et al., 2001; Lengyel et al., 2004; Pissimanis et al., 2000]. Different subregions have, however, been identified where ozone concentration exhibited characteristic spatial and temporal patterns based on the differences arising from the interaction of their respective meteorological conditions with anthropogenic effects [Alvarez et al., 2000]. More specifically,Bloomfield et al., (1996) established a nonlinear regression model for hourly ozone data in the Chicago area in which meteorological variables, seasonality, and a trend term were all implicated. Cox and Chu (1992), on their part, proposed a model for the daily maxima of hourly ozone concentrations based on the Weibull distribution in which the scale parameter is allowed to vary as a function of meteorological conditions. On the other hand, Menezes and Shively (2001) used a multivariate approach to estimate the long-term trend in the extreme values of tropospheric ozone in Houston and Texas. They found that there is a downward trend in the probability of an exceedance followed by a relative flat trend. Shively and Sager (1999) extended the work of Cox and Chu (1992) as well asBloomfield et al., (1996) by using nonparametric regression models to model ozone. The use of multivariate methods was further supported by Lengyel et al., (2004) who analyzed air quality data of which the hidden structure was uncovered by factor analysis and modeled ozone concentration using MLR, PLS, and PCR. While PCA is a very useful tool for selection of properties and different qualities processes leading to a linear model of the data, MLR and PCR or PLS can predict ozone concentration with an error below 2, 5, and 1ppb levels, respectively.

The aim of the present investigation was therefore to analyze environmental data gathered on the daily monitoring of ambient ozone, oxides of nitrogen, and sulfur (IV) oxide at five monitoring sites in Lagos and four monitoring sites in Ilorin, Nigeria. The data were collected between early morning and late evening of the day, and they covered both dry and wet seasons from years 2003 to 2006. The study also included establishment of prediction models on the influence of meteorological parameters on the seasonal variation of concentration of gaseous pollutants in the two Nigerian cities and comparison of measured and modeled concentration values of ozone [Abdul Raheem et al., 2009[b]].

THEORETICAL BACKGROUND

Air Pollution

Air is all around us, odourless, colourless and essential to all life on earth as it acts as a gaseous blanket, protecting the earth from dangerous cosmic radiation from outer space. It helps in sustaining life on earth by screening the dangerous ultraviolet (UV) radiations (< 300nm) from the sun and transmitting only radiations in the range 300nm to 2500nm, comprising of near UV, visible and near infrared (IR) radiations and radio waves (0.01 to 4×10^5nm) [Smart, 1998]. The atmosphere also plays a vital role in maintaining the heat balance on the earth by absorbing the IR radiation received from the sun and re-emitted by the earth. In fact, it is this phenomenon, called "the greenhouse effect", which keeps the earth warm enough to sustain life on the earth. Yet, the air is actually a combination of gaseous elements that have a remarkable uniformity in terms of their contribution to the totality of life. Thus, oxygen supports life on earth; nitrogen is an essential macro - nutrient for plants; and carbon (IV) oxide is essential for photosynthetic activity of plants. Moreover, atmosphere is a carrier of water from the ocean to land, which is so vital for the hydrological cycle. Any major disturbance in the composition of the atmosphere resulting from anthropogenic activities may lead to disastrous consequences or may even endanger the survival of life on earth [Dara, 2004]. The constituent elements are primarily nitrogen and oxygen, with a small amount of argon. Below 100km, the three main gaseous elements, which account for about 99.9% of the total atmosphere, are N_2, O_2 and Ar and they have concentration by volume of 78%, 21%, and 0.93% of respectively [Stanley, 1975]. The presence of trace amounts of other gases would account for the remaining 0.07%. These remaining trace gases exist in small quantities and they are measured in terms of a mixing ratio. This ratio is defined as the number of molecules of the trace gas divided by the total number of molecules present in the volumes sampled. For example, ozone (O_3), carbon (IV) oxide (CO_2), oxides of nitrogen (NO_2 + NO) as NO_x and chloro fluoro carbons (CFCs) are measured in parts per million by volume (ppmv), parts per billion by volume (ppbv) as well as microgram per cubic meter (μgm^{-3}), [Dale, 1976]

Air pollution has been defined [World Bank, 1978] as 'the presence in the outdoor atmosphere of one or more contaminants such as dust, fumes, gas, mist, odour, smoke, or vapour in such quantity, characteristics and duration as to make them actually or potentially injurious to human, plant or animal life, or property, or which unreasonably interferes with the comfortable enjoyment of life and property'. Pollution on the whole is caused principally by human activities, though it can also be a natural process. Air pollution arises from people's economic and domestic activities like modern agriculture which requires agrochemicals. Industrial activities are responsible for wide range of pollution. Thermal power station, burning fossil fuel and moving vehicles emit harmful pollutants like sulphur (IV) oxide, nitrogen (II) oxide and carbon (IV) oxide. Some of these emitted gases have been responsible for acid – rain, global warming and malfunctioning of human / animal's haemoglobin [Stanley, 1975]. Other causes arising from human activities include inappropriate solid waste disposal, gas flaring and oil exploration. Air pollution can also arise from natural causes such as volcanic eruption, whirlwind, earthquake, decay of vegetation, pollen dispersal, as well as forest fire ignition by lightning.

Pollutants

These are substances introduced into the environment in an amount sufficient to cause adverse measurable effects on human beings, animals, plant, vegetation or materials. Pollutants are referred to as primary pollutants, if they exert the harmful effects in the original form in which they enter the atmosphere e.g. CO, NO_x, HC_s, SO_x, particulate matter and so on. On the other hand, secondary pollutants are products of chemical reactions, among primary pollutants are ozone, hydrogen peroxide, peroxyacetylnitate (PAN) and peroxybenzoyl nitrate (PBN). Classification of pollutants can also be according to chemical compositions i.e. organic or inorganic pollutants or according to the state of matter i.e. gaseous or particulate pollutants. Air pollution is basically made up of three components and these are source of pollutants, the transporting medium, which is air and target or receptor which could be man, animal, plant and structural facility.

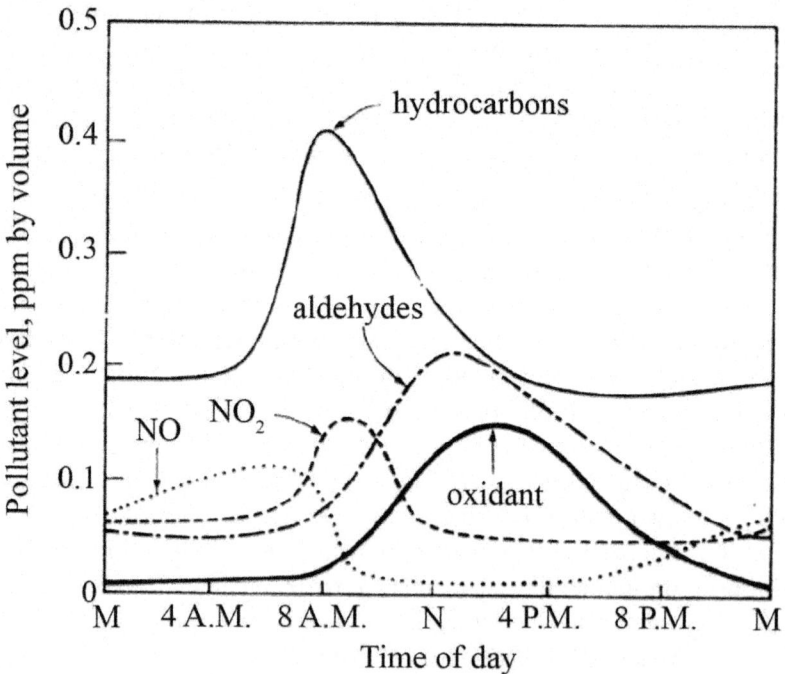

Figure 1. Generalized Plot of Atmospheric Concentrations of Species Involved In Smog Formation As A Function of Time of The Day.

ATMOSPHERIC PHOTOCHEMICAL REACTION

The various chemical and photochemical reactions taking place in the atmosphere mostly depend upon the temperature, composition, humidity and the intensity of sun light. Thus the ultimate fate of chemical species in the atmosphere depends upon these parameters. Photochemical reactions take place in the atmosphere by the absorption of solar radiations in the UV region. Absorption of photons by chemical species gives rise to electronically excited molecules. These reactions are not possible under normal laboratory conditions except at higher temperature and in the presence of chemical catalysts [Hansen *et al*, 1986]. The electronically excited molecules spontaneouslyundergo any one or combination of the following transformations: Reaction with other molecules on collision; Polymerization; Internal rearrangement; Dissociation; De - excitation by fluorescence or De - activation to the original state [Dara, 2004]. Any of these transformation pathways may serve as an initiating chemical step or a primary process. The three steps involved in an overall photochemical reaction are Absorption of radiation, Primary reactions and Secondary reactions.

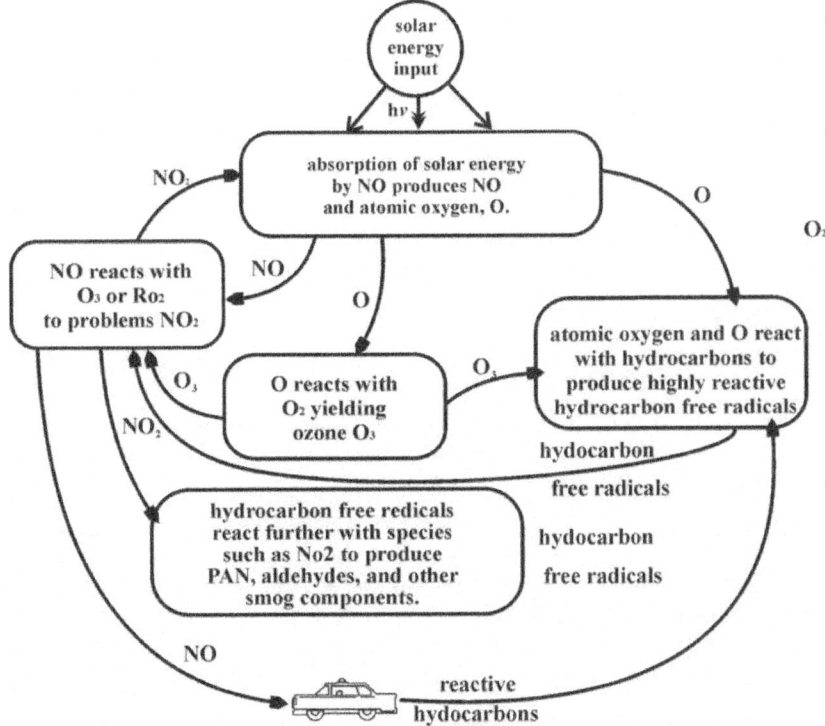

Figure 2. Generalized Scheme For The Formation Of Photochemical Smog.

Smoggy atmosphere show characteristics variations with time of the day in levels of NO, NO_2, hydrocarbons, aldehydes and oxidants. A generalized plot showing these variations is shown in Figure 1. This shows that shortly after dawn the level of NO in the atmosphere decreases markedly, a decrease which is accompanied by a peak in the concentration of NO_2. During the mid – day the levels of aldehydes and oxidants become relatively high, however, the concentration of total hydrocarbons in the atmosphere peaks sharply in the morning, then decreases during the remaining daylight hours. The variations in species concentration shown in the above Figure 1 may be explained by a generalized reaction scheme in Figure 2. This is based on the photochemically initiated reactions which occur in an atmosphere containing oxides of nitrogen, reactive hydrocarbons, and oxygen. The various chemical species that can undergo photo - chemical reactions in the atmosphere include NO_2, SO_2, HNO_3, N_2, ketones, H_2O_2, organic peroxides and several other organic compounds and aerosols; the time – variations of which are explained in a group of overall reactions first proposed by Friedlander nd Seinfeld (1974)

Primary photochemical reaction:

$$NO_2 + h\nu \; NO + O \tag{1}$$

Reactions involving oxygen species

$$O_2 + O + M \; O_3 + M \tag{2}$$

(M is an energy absorbing third body)

$$O_3 + NO \; NO_2 + O \tag{3}$$

Production of organic free radicals from hydrocarbons,

$$RHO + RH \; R \cdot + \text{other products} \tag{4}$$

$$O_3 + RH \; R \cdot + \text{other products} \tag{5}$$

$R \cdot$ is a free radical, which may or may not contain oxygen

Chain propagation, branching and termination

$$NO + R \cdot NO_2 + R^1 \tag{6}$$

In this case $R \cdot$ contains oxygen and oxidizes NO. It is one of many chain propagation reactions, some of which involve NO

$$NO_2 + R \; \text{products (e.g. PAN)} \tag{7}$$

A number of specific reactions are involved in the above overall scheme for the formation of photochemical smog, which is smoke and fog [Thomas *et al*, 1974]. The formation of atomic oxygen by the primary photochemical reaction:

$NO_2 + h\upsilon\ NO + O$

This leads to several reactions involving oxygen and nitrogen oxide species. Examples of such reactions are given below:

$$O + O_2 + M\ O_3 + M \tag{8}$$

$$O + NO + M\ NO_2 + M \tag{9}$$

$$O + NO_2\ NO + O_2 \tag{10}$$

$$O_3 + NO\ NO_2 + O_2 \tag{11}$$

$$O + NO_2 + M\ NO_3 + M \tag{12}$$

$$O_3 + NO_2\ NO_3 + O_2 \tag{13}$$

Table 1. Estimates of USA Primary Pollutants Sources in million tons per year (Million Tons/Year). [Adapted from Thomas *et al*, 1974]

Pollutant Source	CO	NOx	HC	SOx	Particulate	Total
Transportation	111.0	11.7	19.5	1,0	0.7	143.9
Fuel Combustion (Stationary Source)	0.8	10.0	0.6	26.5	6.8	44,7
Industrial Processes	11.4	0.2	5.5	6.0	13.1	36.2.
Solid wastes disposal	7,2	0.4	2.0	0.1	1.4	11.1
Miscellaneous	16.8	0.4	7.1	0.3	3.4	28.0
Total	147.2	22.7	34.7	33.9	25.4	263.9

METHODOLOGY

The determinations of the concentrations of total oxidants (undertaken as O_3), NO_X and SO_2 in the ambient air were carried out between 2003 and 2006 to cover the two seasons; dry season (November to April) and rainy season (May to October) using standard methods. Total oxidants were determined by buffered potassium iodide solution method proposed by Byers and Saltzman (1958). Determination of oxides of nitrogen concentrations were done using the Intersociety Committee Method of Analysis (1972) which is based on the Griess–Saltzman (1954) colorimetric, azo dye forming reagent while oxides of sulphur were determined by conductivity measurements as proposed by Stanley (1975). The application of these techniques to monitoring required the preparation of the following range of reagents as sampling and absorbing solutions as well as preparation of calibration curves. All the chemicals used in the preparation of sampling and absorbing solutions were of analytical grade. The reagents had been prepared and seen stable for months prior to their use. The validation of all procedures was further confirmed using LaMotte standard air sampler.

Ozone

The absorbing solution used for trapping ozone was 1% KI buffered at 6.8 ± 0.2. The standard solution and calibration curve were prepared as follows: 4.09ml of standard 0.001M iodine solution was taken and diluted to 100ml with distilled water. 10ml of this solution was taken and further diluted to 100ml with absorbing reagent and this was labelled solution A. 1ml of A was further diluted to 100ml using absorbing reagent, which gives solution B. This provides a calibrated iodine solution equivalent to 1.92µg of ozone per ml [Stanley, 1975 and Dara, 2004] as ozone reacts with iodine ion in neutral buffer solution according to the following equation:

$$O_3 + 3I^- + H_2O \rightarrow I_3^- + 2\,OH^- + O_2 \qquad (14)$$

The tri – iodide ion liberated has an intense yellow colour. The standard solution was always prepared freshly when needed. A series of standard solutions prepared from above were used to obtain calibration curve. The absorbance measurement was carried out at 352nm. The calibration curve is shown in Figure 3 and it has regression value of 0.9972

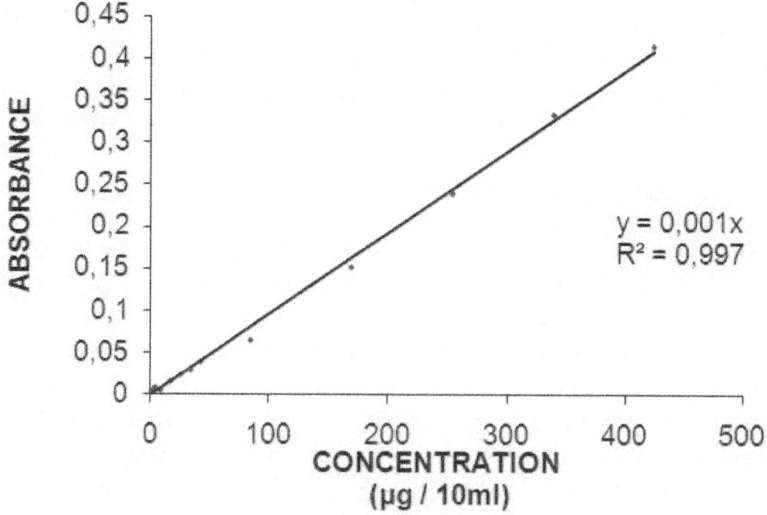

Figure 3. Total oxidants calibration curve.

Sampling Procedure

The set-up of the high volume sampler is given in Figure 4. To follow strict guidelines needed when monitoring for criteria pollutants as discussed earlier, our equipment is validated with LaMotte air sampler (Figure 5) purchased from LaMotte & Company, USA, for capability, repeatability and reliability needed to collect accurate data, and operation of the equipment within our established methods. The impinger used is a big boiling tube of capacity $250cm^3$ with 42mm diameter.

Figure 4. Picture of The Sampling train.

Teflon tubing's used as delivery tubes along with glass tubing that serves as inlet for the ambient air sampled. Silicone grease was used to make the set up airtight by it application to all necessary joints. The air was sampled at the rate of 1 dm^3 min^{-1} with absorbing solution fixed at 30 cm^3 as found appropriate and used [Abdul Raheem et al., 2009[c]], after each sampling for one hour, the impinger was carefully removed and the sample transferred quantitatively into the sample bottle for analysis. The impinger was wrapped with aluminum foil to avoid sunrays' interference. In order to determine pollutant variability over daytime periods, air samples were taken at each location over ten defined sixty - minute periods for any sampling day.

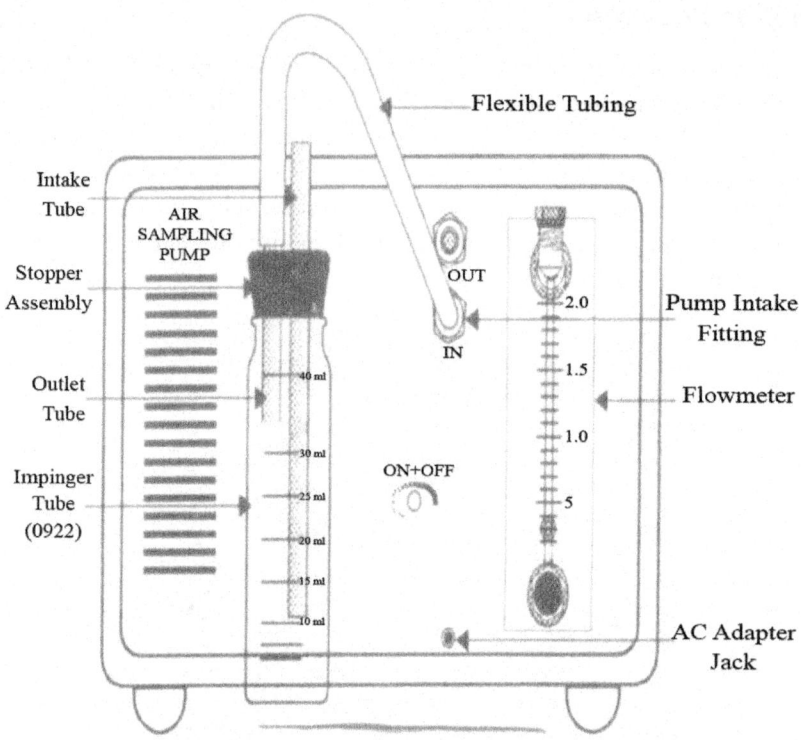

Figure 5. Lamotte Air- Sampler.

The ten sampling periods were spaced equally between approximately 6 am and 6 pm to reflect morning, afternoon and evening. Minimum of 30 samples were collected at each site per week for the pollutants monitored. The time of sixty minutes was found to be optima from the preliminary investigations for the quantitative sampling of these toxic gases within the environment [Abdul Raheem et al., 2009[c]]. All samplings reported were carried out in triplicates. As the road traffic is the common source of pollution cutting across all sites classes, the average traffic volume were determined for all sampling zones. Traffic count was manually done, counting the vehicles passing on the road for 10 minutes in every hour from which hourly traffic was calculated [Abam and Unachukwu, 2009]. The daily minimum and maximum temperatures were between 23 C and 36.5 C, throughout the sampling period.

Analysis

A freshly prepared absorbing solution serves as sample reference or blank solution in order to take care of any impurities during preparation. Absorbance of samples for total oxidants was measured at 352 nm with UV / Visible spectrophotometer. The concentration was read out in μg / 10ml from the reference plot of which one of the examples is shown in Figure 3. The concentrations were converted to μgm^{-3} or ppm or ppb using appropriate conversion factor.

Calculation [Vowels and Connell, 1980]

$$\text{Oxidant} \left(\mu \text{gm}^{-3} \right) = \frac{\text{total } \mu \text{g O}_3 \text{per 10ml of absorbing reagent}}{\text{Volume of air sampled in cubic metres}} \quad (15)$$

$$O_X = \frac{M_{O_3}}{V} \quad (16)$$

$$\mu gm^{-3} = \frac{ppb \times molar\ mass}{24.45} \times 1000 \quad (17)$$

for 1 μgm^{-3} of ozone, the ppb value will be

$$\frac{1 \times 24.45}{48 \times 1000} = 0.51 ppb \quad (18)$$

Table 2. LaMotte total oxidants in air test kit code 7738. Total oxidants in air calibration chart** [LaMotte 6.05] Comparator index number. ** Values in ppm.

QUANTITY	CONTENTS	CODE
2 × 120 Ml	Total oxidants reagent #1	7740-J
30Ml	Total oxidants reagent #2	7741-G
30Ml	Total oxidants reagent #3	7742-G
3	Test Tubes, 5mL, w/ caps	0230
1	Total oxidants in Air Comparator	7739

La Motte Total Oxidants Sampling Procedure

10 mL of reagent #1 was put into impinging tube, followed by 2 drops of reagent #2 added and swirled to mix then 2 drops of reagent #3 added and also swirled to mix. The impinging apparatus was connected to intake of the sampling pump as shown in Figure 5 such that the long tube was immersed in the absorbing solution. The impinging tube was covered with foil to protect it from light while sampling. The flow meter of sampling apparatus was adjusted

to collect air at 1.0 Lm^{-1} rate. The sampling continued until 15 minutes when a measurable pink colour developed. The impinging tube was disconnected from the pumping apparatus and the contents poured into a clean test tube (0230). The test tube was later inserted into the total oxidants in air comparator (7739) and the sample colour was matched with an index value. The index value was recorded and the calibration chat was used to convert the index readings into concentration of the pollutant in the atmosphere in parts per million.

Table 3. Total oxidants in air calibration chart** [LaMotte 6.05] Comparator index number. ** Values in ppm.

Time(min)	1	2	3	4	5	6	7	8
5	0.14	0.36	0.72	1.08	1.44	2.88	4.32	5.76
10	0.07	0.18	0.36	0.54	0.72	1.44	2.16	2.88
15	0.05	0.12	0.24	0.36	0.48	0.96	1.44	1.72

Oxides of Nitrogen (NO$_x$)

The absorbing solution used for trapping NO$_x$ was Saltzman solution which is an azo dye forming reagent.

The standard solution and calibration curve were prepared as follows: 2.16g of sodium di - oxo nitrate (III), $NaNO_2$ was dissolved in 1000 cm^3 volumetric flask and the solution labeled A. 1ml of solution A was measured out into 100 ml volumetric flask and the solution made up to the mark. This solution of concentration 0.0216 gL^{-1} was labeled B. 1ml of B was added to 100 ml volumetric flask and distilled water added to the mark to give 0.000216 gL^{-1} solution C. 10ml of solution C was added to 100ml volumetric flask and filled to the mark with distilled water to give solution of concentration 0.0000216 gL^{-1} labeled D. Further dilutions of the last two solutions C and D were used for calibration plot. As the standardization was based on the empirical observation that 0.72 mole of $NaNO_2$ produces the same colour as 1mole of NO$_2$ [Hesketh, 1972]. In other words, 1ml of the 0.000216 gL^{-1} working standard which contains 0.216µg of $NaNO_2$ should be equivalent to 0.2 µg of NO$_2$ Series of standard solutions prepared in 10 ml volumetric flasks from solutions C and D above were allowed to stay for 15 minutes for colour development and the spectra run at 550 nm to obtain a set of absorbance value which were recorded against known concentrations. The formation of red azo dye of which the absorbance is picked at 550 nm can be explained according to the equation in Figure 6 However, a plot of absorbance against concentration in µg / 10 ml was made, a straight line graph obtained with regression value of 0.9962 as shown in Figure 7.

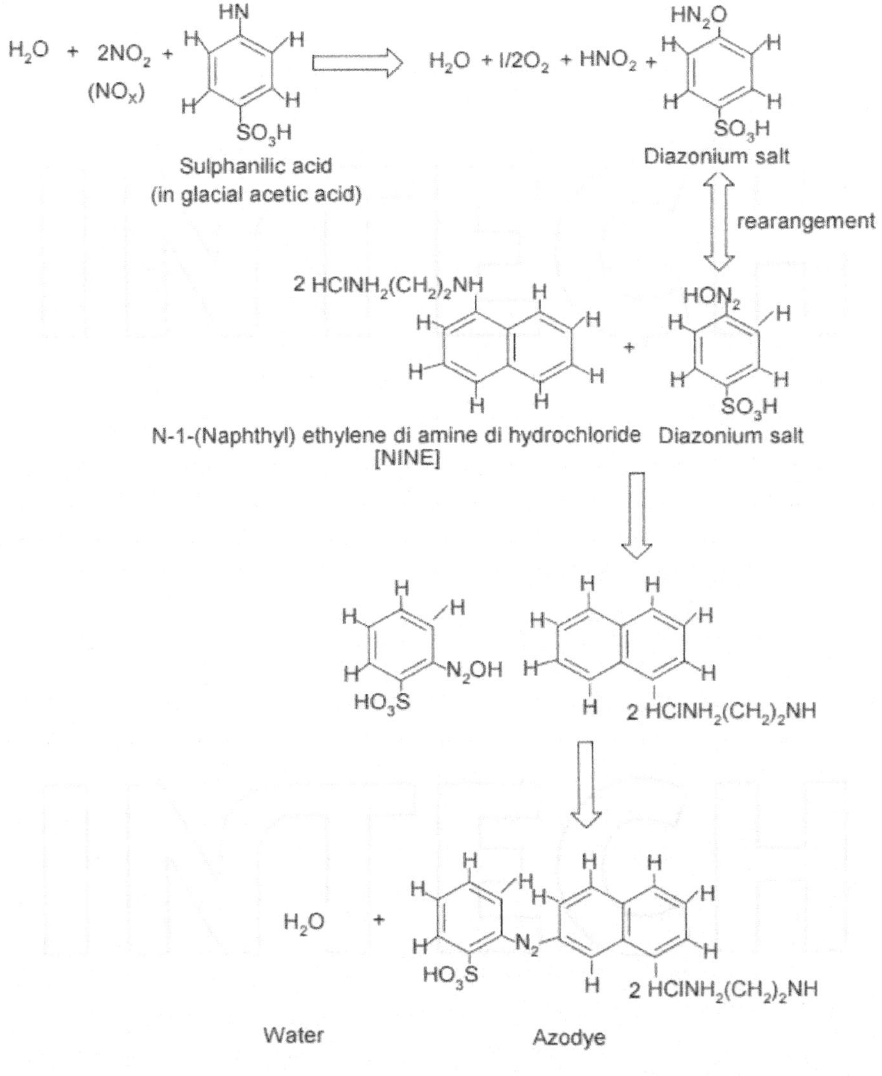

Figure 6. Equation showing the formation of azo dye.

Sampling Procedure

The procedure for sampling is as given above

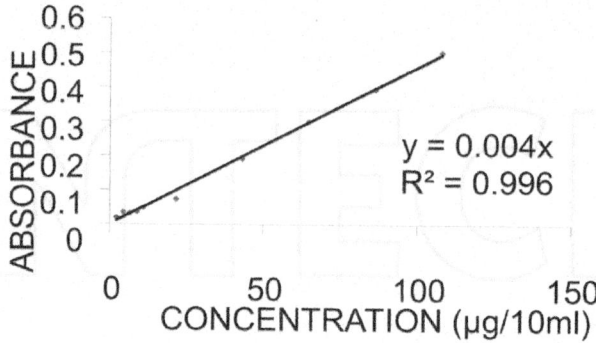

Figure 7. Oxides of nitrogen (no$_x$) calibration curve.

Analysis

The absorbing solution serves as sample reference or blank solution in order to take care of any impurities during preparation. Absorbance of samples for oxides of nitrogen was measured at 550nm with UV / Visible spectrophotometer. The concentration was read out in µg / 10ml from the reference plot of which an example is shown in figure 7. The concentrations were converted to µgm^{-3} or ppm or ppb of which the conversion factors are explained hereafter.

Calculation [Vowels and Connell, 1980]

$$NO_x\left(\mu gm^{-3}\right) = \frac{total \ \mu g \ NO_x \ per \ 10ml \ of \ absorbing \ reagent}{Volume \ of \ air \ sampled \ in \ cubic \ metres}$$

(19)

$$NO_x = \frac{M_{NO_x}}{V}$$

(20)

$$\mu gm^{-3} = \frac{ppb \times molar \ mass \times 1000}{24.45}$$

(21)

for 1 µgm^{-3} of NO$_x$ as NO$_2$, the ppb value will be

$$\frac{1 \times 24.45}{46 \times 1000} = 0.53 ppb$$

(22)

Table 4. LaMotte nitrogen (IV) oxide in air test kit code 7690

QUANTITY	CONTENTS	CODE
2 × 120 mL	Nitrogen (Iv) oxide reagent #1 Absorbing solution	7684-J
30 mL	Nitrogen (Iv) oxide reagent #2	7685-G
10g	Nitrogen (Iv) oxide reagent #3 powder	7688-D
2	Test tubes,10mL, glass, w/caps	0822
1	Spoon, 0.005g, plastic	0696
1	Pipet, droping, plastic	0352
1	Nitroge (IV) oxide in air comparator	7689
1	Tubing	23609
1	Pipet	30410
1	Needle	27336-01

Nitrogen (IV) Oxide Lamotte Sampling Procedure

10mL of reagent #1 i.e. absorbing reagent was poured into the impinging tube, a gas bubbler impinger (0934). The impinging apparatus was connected to the intake of air sampling pump and the long tube was immersed in the absorbing solution. The special adaptor was attached to the intake of the pump to sample at $0.2Lm^{-1}$ while the sampling was done for 20 minutes when a measurable amount of nitrogen (IV) oxide was absorbed. At the end of the sampling period the contents of the impinging tube was poured into test tube (0822). The pipette (0352) was used to add a drop of reagent #2, the test tube capped and mixed after which the 0.05g spoon was used to add 0.05g of reagent #3. The test tube capped and the solution left for 10 minutes for colour development after which the test tube was placed into comparator (7689) and the sample colour matched to index of colour standards. The index number which gave the proper colour matched was recorded and the calibration chart used to convert the index read to concentration of nitrogen (IV) oxide in ppm.

Table 5. Nitrogen (IV) oxide in air calibration chart

	Comparator index number							
Time (min)	1	2	3	4	5	6	7	8
1	0.00	2.8	7.0	14.0	21.0	28.0	42.0	56.0
5	0.00	0.56	1.40	2.80	4.20	5.60	8.40	11.20
10	0.00	0.28	0.70	1.40	2.10	2.80	4.20	5.60

| 15 | | 0.00 | 0.19 | 0.47 | 0.93 | 1.40 | 1.87 | 2.80 | 3.74 |
| 20 | | 0.00 | 0.14 | 0.35 | 0.70 | 1.05 | 1.40 | 2.10 | 2.80 |

Sulphur (IV) Oxide

The absorbing solution used for trapping SO_2 was 0.3M H_2O_2 solution buffered at pH 5 ± 0.2.

The standard solution and calibration curve were prepared as follows:

0.1M H_2SO_4 was used as parent standard solution. All other lower concentrations were prepared from serial dilution of 0.1M H_2SO_4. 0.1M H_2SO_4 was standardized by titration against Na_2CO_3 using methyl orange as indicator. The conductivity measurement of each of the concentrations of H_2SO_4(0.001 – 0.01M) obtained from serial dilution were taken, using Hanna Instrument EC 214 conductivity model. A graph of conductivity values in Siemens per centimeter (Scm^{-1}) against concentrations of H_2SO_4 in mol dm^{-3} was plotted. The data gave a straight line which passes through the origin with regression value of 0.9874. The calibration curve so obtained is shown in Figure 8. This was used as a working curve for the determination of SO_2 during the analysis of samples.

Sampling Procedure

The procedure for sampling others remained except the flow rate that was increased to 2 Lmin^{-1} for optimization purpose [Abdul Raheem et al., 2009^c].

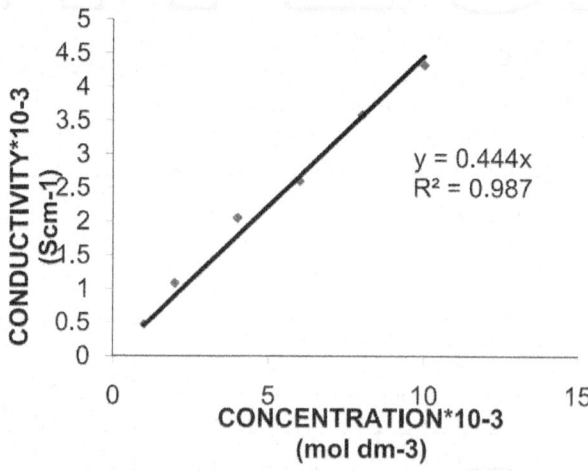

Figure 8. Sulphur (iv) oxide calibration curve.

Analysis

Conductivity measurements were undertaken using the Hanna Instrument Model E 214 conductivity meter.From the sample and reference solutions 20 cm³ volume was measured respectively into a liquid sample holder test tube of Hanna model conductivity meter. The concentrations in mol dm⁻³ of H_2SO_4 formed from SO_2 of the samples were read out from the reference plot (Fig.8), the concentrations obtained in mol dm⁻³ were converted to parts per million or parts per billion or microgram per cubic meter (ppm or ppb or μgm⁻³) as shown below using appropriate conversion factor. Equation of reaction for formation of H_2SO_4 from SO_2 is shown below:

$$SO_2 + H_2O_2 \rightarrow H_2SO_4$$

(23)

Calculation [Stanley, 1975; Vowels and Connell, 1980]

$$ppm = \frac{moldm^{-3} \times mmSO_2 \times samplingvol.}{flowrate \times samplingduration}$$

(24)

$$\mu gm^{-3} = \frac{ppm \times molar\,mass}{24.45} \times 1000$$

(25)

for 1 moldm⁻³ of SO_2 ppm value will be

$$= \frac{1 \times 64 \times 30}{2 \times 60 \times 1000}$$

$$= 1.6 \times 10^{-2} \, ppm$$

(26)

in μgm⁻³, the value becomes:

$$\frac{1.6 \times 10^{-2} \times 64 \times 1000}{24.45} = 41.88 \mu gm^{-3}$$

(27)

Table 6. LaMotte sulphur (IV) oxide in air test kit code 7714

QUANTITY	CONTENTS	CODE
2 × 250 mL	Sulphur (IV) oxide absorbing solution	7804-K
15g	Sulphur (IV) oxide reagent #1	7693-E
30mL	Sodim hydroxide, 1.0 N	4004PS-G
60mL	Sulphur (IV) oxide passive bubbler indicator	7805-H
2	Pipets, 1.0mL, plastic	0354
2	Test tubes, 5 mL, plastic, w/caps	0230
2	Test tubes, Hester, w/caps	0204
1	Spoon, 0.25g	0695
1	Dispenser caps	0693
1	Sulphur (IV) oxide passive bubbler comparator	7746

Sulphur (IV) Oxide Lamotte Sampling Procedure

10mL of Sulphur (IV) oxide absorbing solution was added to impinging tube and connected to the impinging apparatus as shown in Figure 5. The long tube was immersed into the absorbing solution. Sampling was done at 1.0 Lpm for 60 minutes or 90 minutes. The impinging apparatus was covered with foil to protect it from light. At the end of the sampling time the small test tube (0230) was filled to the line with the sample and 0.25g spoon was used to add a level measured of Sulphur (IV) oxide reagent #1. The test tube containing the mixture was capped and vigorously shaken to dissolve the powder. A 1 mL pipette was used to add 1mL sodium hydroxide, 1.0N, to the same small test tube, capped and inverted several times to mix. The other 1mL pipette was also used to add 2mL (2 measures) of Sulphur (IV) oxide passive bubbler indicator (7805) to a large test tube (0204). The contents of the small test tube were poured into the large test tube containing the indicator. Immediately the tube capped and inverted six times, holding the cap firmly in place with the index finger. After waiting for 15 minutes, the test tube was placed into the Sulphur (IV) oxide passive bubbler comparator (7746). The sample colour matched with the standard colour and the index number read and recorded from the comparator. The index number was converted to concentration in ppm using the calibration chart provided.

Table 7. Sulphur (IV) oxide in air calibration chart.** Values in ppm.

Time (min)	Comparator index number							
	1	2	3	4	5	6	7	8
10	0.00	0.19	0.29	0.38	0.48	0.57	0.67	0.76
30	0.00	0.06	0.10	0.13	0.16	0.19	0.22	0.25
60	0.00	0.03	0.05	0.06	0.08	0.10	0.11	0.13
90	0.00	0.02	0.03	0.04	0.05	0.06	0.07	0.08

QUALITY ASSURANCE

The impinger was well rinsed with distilled water and properly wrapped with foil paper before each use. The tubing's and corks in the sampling train were checked before and during sampling, in case they had become slackened, however silicone grease was used to increase the pressure by making them air tight.

The absorbing reagents were always prepared freshly ahead of sampling for the solution to stabilize. They were stored in amber coloured bottles and refrigerated because of light interference. They were always allowed to thaw and assume the 25 C temperature before use.

Lengthy contact with air by the absorbing reagent was avoided during both preparation and use to prevent absorption of the oxides. The absorbance of the reagent blank was deducted from that of the samples where the machine could not be adjusted to zero to avoid matrix error, especially with the conductivity meter.

For the nitrogen oxides determination, a gas bubbler impinger (fritted gas bubbler) was used instead of a general purpose impinger as absorption tube. The general purpose impinger has been reported to give low absorption efficiency with oxides of nitrogen [ICMA, 1972; Onianwa et al., 2001; Saltzman, 1954]. However the results were corrected and correlated with the fritted bubbler as well as standardized absorbing solution imported from LaMotte and Company, USA.

Greatest accuracy has been reported to be achieved by standardizing the sampling train with accurately known gas sample in a precision flow dilution system like a permeation tube [Dara, 2004]. Due to lack of the apparatus necessary for the standardization of the train, the actual collection efficiency is not known. However with the use of LaMotte sampling pump with inbuilt flow meter and standardized reagents, we recorded high collection efficiency at sites with increase concentrations of samples.

RESULTS

This is already discussed extensively in Abdul Raheem, 2007 and Abdul Raheem et al., 2009[a,b,c]. Typical tables are shown to show the typical measurements concentration results and the meteorological data

Table 8. Dry season environmental data for Ilorin

Start of sampling	End of sampling	OX (ppb)	NOx (ppb)	SO2 (ppb)	REL-HUM (%)	WND ms-1	DWND (oC)	AIR-TEMP (oC)	Sun Exp Wm-2
6.30am	7.30am	29.08 ±11.73	1.47	7.83	78.17	27.60	144.60	22.70	-1.55
7.45am	8.45am	29.72 ±10.5	3.44	6.54	71.67	36.30	156.40	23.20	0.51
9.00am	10.0am	29.71 ±5.57	0.43	4.17	57.30	44.60	156.50	27.90	8.63
10.15am	11.15am	33.11 ±5.51	1.67	4.42	53.30	42.00	160.50	29.80	12.61
11.30am	12.30pm	46.69 ±7.49	1.73	6.27	42.00	42.30	153.20	31.50	15.36
12.45pm	1.45pm	69.94 ±15.45	1.04	7.36	38.67	43.40	154.00	32.80	16.09
2.00pm	3.00pm	35.55 ±11.21	2.46	8.84	35.50	41.60	160.00	34.30	13.39
3.15pm	4.15pm	21.44 ±6.31	2.46	7.62	37.17	39.40	167.90	33.80	10.16
4.30pm	5.30pm	17.62 ±3.13	2.69	9.52	39.00	39.30	178.00	33.00	5.66
5.45pm	6.45pm	11.56 ±2.19	2.91	9.11	42.67	37.60	176.70	31.30	0.86

Table 9. Dry season environmental data for Lagos

Start of sampling	End of sampling	OX (ppb)	NOx (ppb)	SO2 (ppb)	REL-HUM (%)	WND ms-1	DWND (oC)	AIR-TEMP (oC)	Sun Exp Wm-2
6.30am	7.30am	14.26	12.40	10.72	90.58	3.07	107.02	25.99	-1.65
7.45am	8.45am	22.92	5.89	7.20	87.58	4.67	156.38	26.38	2.80
9.00am	10.0am	28.95	5.39	11.15	73.75	7.09	189.02	29.05	8.90
10.15am	10.0am	46.86	5.66	14.82	67.92	7.35	182.63	30.25	11.00
11.30am	12.30pm	43.21	6.41	10.51	63.50	8.76	170.67	31.30	12.06
12.45pm	1.45pm	85.31	5.68	12.74	60.33	10.11	159.55	32.00	17.30
2.00pm	3.00pm	73.77	6.45	16.62	60.08	10.36	155.00	31.98	15.10

3.15pm	4.15pm	26.06	6.84	15.47	62.67	10.94	163.79	31.37	13.20
4.30pm	5.30pm	12.23	5.72	16.48	67.00	10.21	165.22	30.38	10.70
5.45pm	6.45pm	8.58	6.90	19.21	72.75	8.99	166.64	29.20	3.30

This is showing typical results of statistical modeled analysis of Ilorin and Lagos during dry season MLR with backward selection in stepwise mode (without intercept) results in the following equation:

$$OX_{ILO} = 6.092 \times SO_2 + 0.657 \times RHUM - 2.653 \times ATEMP + 4.385 \times SUNEXP \qquad (28)$$

Where $R = 0.981$, $F (4, 6) = 38.389$, $p < 0.000$

This shows that only four of the variables are found to be significant for retention in the model.

MLR using backward selection in stepwise mode (without intercept) results in the following equation:

$$OX_{Lag} = 1.679 \times ATEMP + 5.622 \times SUNEXP - 8.079 \times WND \qquad (29)$$

where, $R = 0.961$, $F (3, 7) = 27.874$, $p < 0.000$

MLR shows that only three of the variables are significant for retention in the model.

A table Comparing the ozone measured concentration with calculated results from MLR model equations

Table 10. MLR equation modeled results for ozone compared with monitored results for the two cities of interest during rainy and dry seasons (ppb)

	ILORIN		LAGOS	
	RAIN	DRY	RAIN	DRY
MEASURED	21.86 ± 2.47	32.44 ± 5.13	9.87 ± 0.99	36.22 ± 5.76
MODELED	16.12 ± 1.86	44.32 ± 4.25	9.89 ± 0.82	36.29 ± 3.87

GENERAL CONCLUSION

The direction and spatial extent of transport and the relative contribution of transported ozone and precursors to individual downwind areas are highly variable. A number of factors influence site to site differences in ozone concentrations, including sources of precursor's emissions and meteorological conditions.

Data analysis also reveals that NO_x and SO_2 as well as volatile organic compounds contribute to ozone formation and this is in accordance with other researchers [Winer *et al*, 1974; Canada – US, 1999; chou *et al*, 2006]. The relative effectiveness of reductions of these three precursors can vary with location and atmospheric condition. Overall the concentrations of ozone could be said to be influenced globally by background concentrations, locally generated concentrations and transported concentrations.

On the whole the chemometric multivariate analysis results confirmed our experimental results and unfold the fact that meteorological influence plays a major role in the atmospheric chemistry of ozone.

Finally, these results and analysis suggested that ozone acting in concert with other pollutants need to be recognized as important health and ecosystem related air quality concern in Nigeria. Based on increasing evidence on regional transport of ozone all over the world, there is need for recognition that ground – level ozone would be an appropriate issue to be considered by the Nigerian government. In particular, a proactive measure has to be formulated towards reducing NO_x and SO_2 and by consequence O_3 in Nigeria.

REFERENCES

1. Raheem. A. M. O. Abdul, F. A. Adekola, I. B. Obioh, 2009 Bull. Chem. Soc. Ethiop., 23 (3), 383390

2. Raheem. A. M. O. Abdul, F. A. Adekola, I. B. Obioh, 2009 Environ. Model. Assess. 14 487509 b

3. Raheem. A. M. O. Abdul, F. A. Adekola, I. B. Obioh, 2009 SCIENCE FOCUS, 14 2 166185 c

4. Raheem. A. M. O. Abdul, 2007 Ph.D. Thesis, University of Ilorin, Nigeria

5. ACGIH 1991 American Conference of Governmental Industrial Hygienists Documentation of Threshold Limit Values and Biological Exposure Indices, 2 786788 . 6th ed., ACGIH Cincinnati

6. E. Alvarez, F. Pablo, C. Thomas, S. Rivas, 2000 International Journal of Biometeorology, 44 4451 .

7. G. A. Bakken, D. R. Long, J. H. Kalvis, 1997 Examination criteria for local model principal component regression. Applied Spectroscopy 51 18141822 .

8. P. Bloomfield, J. A. Royle, L. J. Steinberg, Q. Yang, 1996 Atmospheric Environment 30, 3067

9. C. C. Chou, K. , S. C. Liu, C. Lin, Y. , C. Shiu, J. , K. Chang, H. , 2006 Atmospheric Environment 40 38983908 .

10. W. M. Cox, S. H. Chu, 1992 Atmospheric Environment, 27B, 425.

11. S. S. Dara, 2004 A Textbook of Environmental Chemistry and Pollution Control, S. Chand and Company Ltd. New Delhi 110055

12. ECE 1984 Effects of acidifying depositions and related pollutants on forest ecosystems, Executive Body for the Convention on Long-Range Transboundary Air Pollution, Third Session, Geneva; 59 March.

13. J. W. Einax, H. W. Zwanziger, S. Geiss, 1997 Chemometrics in environmental analysis. Weinheim: Wiley

14. J. E. Jackson, 1991 A user's guide to principal components New York: Wiley

15. D. A. T. Jaffe, D. Covert, R. Kotchenruther, B. Trost, J. Danielson, W. Simpson, 1999 Geophysical Research Letters 26 711714 .

16. D. Klaus, A. Poth, M. Voss, 2001 Atmosfera 14 171188 .

17. A. Lengyel, K. Heberger, L. Paksy, O. Banhidi, R. Rajko, 2004 Chemosphere 57 889896 .

18. K. A. Menezes, T. S. Shively, 2001 Environmental Science & Technology 35 25542561 .

19. C. Norman, 1987 Analyzing multivariate data. San Diego, CA: Harcourt Brace Jovanovich

20. P. C. Onianwa, S. O. Fakayode, B. O. Agboola, 2001 Bull. Chem. Soc. Ethiop. 15, 71

21. M. Otto, 1999 Chemometrics statistics and computer application in analytical chemistry. Weinheim: Wiley-VCH

22. D. K. Pissimanis, V. A. Notaridou, N. A. Kaltsounidis, P. S. Viglas, 2000 Theoretical and Applied Climatology 65 4962

23. P. Schutt, E. B. Cowling, Disease. Plant, 1985 69, 548.

24. T. S. Shively, T. W. Sager, 1999 Environmental Science & Rechnology 33, 3873

25. N. H. Timm, 1985 Applied multivariate analysis. Englewood Cliffs, NJ: Prentice-Hill

26. R. Vautard, P. H. J. Builtjes, P. Thunis, C. Cuvelier, M. Bedogni, B. Bessagnet, 2007 Atmospheric Environment 41 173188

27. N. B. Vogt, 1989 Chemometrics and Intelligent Laboratory Systems 7 119130

28. J. G. J. Watson, 1984 Air Pollution Control Association 34 619623

29. World Health Organization (WHO) 1981 Sulfur dioxide, Environmental Health Criteria, WHO: Geneva

Chapter 3

IMPACT ANALYSIS OF AIR POLLUTANT EMISSION POLICIES ON THERMAL COAL SUPPLY CHAIN ENTERPRISES IN CHINA

Xiaopeng Guo, Xiaodan Guo, and Jiahai Yuan

School of Economics and Management, North China Electric Power University, Hui Long Guan, Chang Ping District, Beijing 102206, China

ABSTRACT

Spurred by the increasingly serious air pollution problem, the Chinese government has launched a series of policies to put forward specific measures of power structure adjustment and the control objectives of air pollution and coal consumption. Other policies pointed out that the coal resources regional blockades will be broken by improving transportation networks and constructing new logistics nodes. Thermal power takes the largest part of China's total installed power generation capacity, so these policies will undoubtedly impact thermal coal supply chain member enterprises. Based on the actual situation in China, this paper figures out how the member enterprises adjust their business decisions to satisfy the requirements of air pollution prevention and control policies by establishing system dynamic models of policy impact transfer. These dynamic analyses can help coal enterprises and thermal power enterprises do strategic environmental assessments and find directions of sustainable development. Furthermore, the policy simulated results of this paper provide the Chinese government with suggestions for policy-making to make sure that the energy conservation and emission reduction policies and sustainable energy policies can work more efficiently.

INTRODUCTION

According to the Air Quality Report issued by the Chinese Environmental Protection Ministry, in 2013 China's annual haze days reached the highest level on record (19.5 days) owing to unsustainable development and an unreasonable energy structure. In addition, the national annual concentration of PM2.5 has reached 72 micrograms per cubic meter. Only three cities reached the standard, which accounts for 4.1 percent of all 74 cities. The Intergovernmental Panel on Climate Change (IPCC) found that coal combustion is the main source of air pollutants. In addition, the research done by the Chinese Academy of Sciences (CAS) showed that: the sources of Beijing's particulate matter (PM10) emission are secondary inorganic aerosols (26%), industrial pollution (25%), coal consumption (18%), soil dust (15%), biomass burning (12%), and automobile exhaust gas and waste incineration (4%). The secondary inorganic aerosols, industrial pollution, and coal consumption are all caused by the burning of fossil fuels. Therefore, China's worsening air quality is probably caused by extensive coal consumption and a high energy consumption development pattern.

Coal consumption is considered to be the second most important anthropogenic contributor to global air pollution [1,2]. From the perspective of coal consumption structure, the electric power industry is the biggest contributor to the coal resources consumption. The power generation industry produces more than 40 percent of total air pollution emissions in China [3]. China's past electricity development policies have led the power industry to make excessive investment in thermal power installed capacity and low operational efficiency [4]. If a long-term transition towards a low carbon economy is not carried out, China's CO_2 emissions could rise by 160%–250% from 2010 to 2050 [5]. By the year 2020, China's thermal power generation may reach over 7 trillion kilowatt-hours, and air-pollution intensity will be nearly twice the 2005 level [6].

To accelerate the control of air pollution, the Chinese government has introduced many policies managing the air quality and adjusting the energy structure since 2009. These air pollution emission policies will surely change the operating environment of coal supply chain enterprises. So it is urgent to analyze the policy impact and optimize the development patterns for coal and power industries under new policy restraints [7]. From the perspective of the power industry, coal-fired power will translate into clean and efficient energy power after the implementation of sustainable development policies [5,8]. China's hydropower, wind power, and nuclear power industry will meet a tremendous need in the following decades under the encouragement of energy structure adjustment and an emission reduction policy [9,10]. Nuclear and hydropower may play a dominant role in contributing to China's air pollution reduction in the long term [11]. Also, the promotion of renewable energy utilization will surely have a great effect on China's coal and power industry [12]. Further, the air pollution emission reduction policies can also accelerate the technological advancement and clean coal utilization of thermal coal supply chain enterprises [13,14].

The thermal coal supply chain system is the whole system of coal enterprises, power enterprises, and coal transportation enterprises, which guide the process from coal production to coal consumption. In addition, air pollutant emission reduction policies' impact on the thermal coal supply chain is a complex, dynamic evolution process concerning many fields such as policies related to energy conservation and emission reduction, economic development, power production and consumption, resource exploitation and utilization, and energy price. So it is obvious that the power structure adjustment has nonlinear characteristics. The system dynamics (SD) method not only models the market's real behavior but also properly explains the relationship between the main variables of the system [15]. Considering the advantages of integrity and dynamics that system dynamics has in analyzing complex dynamic problems, this paper set up a complete system dynamics model by analyzing: the air pollution emitted during coal combustion, coal washing technology, installed capacity, unit transform, and new energy power generation, under the constraint of the new atmospheric pollutant emission policy, to seek a development pattern for the thermal coal supply chain.

The SD approach has been applied in investigating the sustainable management of electric power systems. Some scholars have set up an SD-based model to investigate the distributed energy resource expansion planning [15,16,17] and energy efficiency improvement [18], considering both energy states and production constraints. Other scholars use SD methodology to simulate the behavior of the renewable energy sectors such as nuclear [19] and photovoltaic energy [20]. SD models are also widely built to explore the effects of energy consumption and CO_2 emission reduction policies [21,22,23,24]. Previous models have structured the investment, dispatch, pricing heuristics, and electricity generation resource factors with common emission reduction policies such as feed-in-tariffs, investment subsidies, and carbon taxes. In order to deeply investigate the new development pattern of the power and energy industry, some scholars have qualitative explored the link between transportation systems and air pollution reduction policies using the SD approach [25,26]. However, the dynamic behavior of the thermal power system and the complicated feedback of a coal system under pollution reduction policies were not taken into account in previous studies.

As policy refinement increases, China's air pollutant emission reduction policies select the specific goals of pollution emission, coal cleaning proportion, and desulfurated capacity proportion. Under the new policy situation, the former system dynamic model is not suitable to measure the effect on China's thermal coal supply chain member enterprises.

The objectives of this paper are as follows:

- To analyze China's recent air pollution reduction policies systematically.
- To apply some descriptive and effective methodologies to simulate the fundamental and dynamic development process of thermal coal supply chain member enterprises driven by emission reduction policies.
- Taking China as a case study, to develop a system dynamics model based on Vensim PLE for Windows Version 5.10a software (Ventana Systems, Inc., Cambridge, MA, USA), which offers a realistic platform for predicting the development trends of China's thermal coal supply chain enterprises from 2012 to 2022.

The simulated results are significant in predicting energy proportion, power proportion, and development routes of thermal coal supply chain member enterprises. In addition, the simulations of China's air pollution emission reductions have meaning for policy-making institutions.

CHINA'S AIR POLLUTION EMISSION REDUCTION POLICIES

China is the world's biggest energy consumer [27,28]. According to the information distributing platform of National Statistics, China's coal consumption accounts for about 70% of the total energy consumption in 2013. The thermal coal consumption of China's electric industry is 1967 million tons, accounting for 64.29% of the total coal consumption. China's growing atmospheric pollution problem is caused by the long-term accumulation of multiple factors such as unsustainable development patterns and unreasonable energy electric power industry structures. In order to better reduce air pollution, the Chinese government has introduced many laws involving concrete measures since 2009.

In 2009, after realizing the importance of sustainable development the Chinese government first proposed to reduce carbon dioxide emissions per unit of GDP by 40%–45% in 2020, as of the 2005 level. This restrictive goal has also been included in the medium and long term planning of China's national economic and social development.

In 2010, the fifth plenary session of the 17th CPC Central Committee emphasized that China should make the construction of a resource-saving and environment-friendly society an important focus despite the acceleration of economic growth and the readjustment of the economic structure. Prime

Minister Wen Jiabao promised that China would energetically promote energy conservation and raise the efficiency of energy consumption, especially in industry and transportation. Wen also said that China would add 80 million tons of standard coal energy saving ability annually, and that all additional and reconstructive coal-fired units must build and run flue gas desulfurization facilities synchronously.

Since 2012, hazy weather in the Beijing-Tianjin-Hebei and Yangtze River Delta areas is becoming more and more frequent, so the importance of air pollutant prevention is becoming more and more noticeable. In October 2012, the Chinese National Development and Reform Commission, the Ministry of Environmental Protection, and the Ministry of Finance jointly issued The 12th Five-Year Plan for the prevention and control of atmospheric pollution in key areas. The 12th Five-Year Plan pointed out that we should focus on the optimization of industrial layout and energy structure and the enhancement of clean energy under the current serious situation of air pollution. China should adhere to a diversified energy development strategy, strive to improve the proportion of clean low-carbon fossil energy and non-fossil energy, promote the efficient utilization of clean coal, implement the alternatives of traditional energy, and speed up the optimization of energy production and consumption structure. By the end of the 12th period, the proportion of non-fossil energy consumption should increase to 11.4% and the proportion of coal consumption should decrease to 65%. The proportion of non-fossil energy generation installed capacity should reach 30%.

In June 2013, the premier, Li Keqiang, introduced ten measures for the control of air pollution in the state council executive meeting. The main content of these measures is the reduction of air pollutant emission, the control of high energy-consuming enterprises, the adjustment of energy structures, and the new energy conservation and emissions reduction mechanisms of incentive and constraint. We should speed up the adjustment of energy structure by implementing the interregional transmission project, controlling coal consumption reasonably, and promoting the use of clean coal. By the end of 2014, China should complete the elimination of backward installed capacity ahead of time. The Chinese government should follow the guiding and incentive rule of taxes and subsidies, in order to push for sustainable development.

In September 2013, The Action Plan for the Control of Air Pollution, issued by the state council, clearly pointed out that in 2017 the inspirable particle concentrations of China's prefecture-level cities should be reduced by 10% from 2012 levels and the inspirable particle concentrations of Beijing-Tianjin-Hebei, the Yangtze River Delta area, and the Pearl River Delta area should be reduced by about 25%, 20%, and 15%, and the PM10 concentrations of Beijing especially should be reduced to under 60 micrograms per cubic meter. The Action Plan for the Control of Air Pollution also mentioned the adjustment of energy structure, the optimization of industrial layout, the improvement of environmental economic policies, and some other measures. Measures optimizing the industrial layout covered the capacity limitation of energy-intensive and highly polluting industries, the acceleration of backward capacity elimination, and the compression of excessive capacity. China should take a sustainable development pattern by controlling coal consumption, increasing the proportion of washed coal, accelerating the use of clean energy, and raising the energy usage effectiveness. An energy conservation and emissions reduction mechanism of incentive and constraint should be promoted actively in order to improve sustainable environmental economic policies

In addition, some local governments introduced policies to cooperate with the implementation of The Action Plan for the Control of Air Pollution, such as an action plan implementing rules for the control of air pollution in Beijing-Tianjin-Hebei and surrounding areas. These rules made more detailed targets for the provinces and cities of the Beijing-Tianjin-Hebei region. The plan pointed out that by the end of 2017 the proportion of Beijing's coal consumption will drop below 10% and high-quality energy such as electricity and natural gas will account for more than 90%. Beijing will cut 13 million tons of raw coal by using many comprehensive measures such as eliminating backward production capacity, clear violations capacity, strengthening energy conservation and emissions reduction, implementing clean energy replacement, safe development of nuclear power, and strengthening the new energy-efficient utilization.

Overall, these air-pollution reduction policies focused on optimizing the industrial structure and energy structure, and giving impetus to industrial transformation and upgrading. Along with the restriction of coal consumption and the development of clean energy generation, the business environment and development modes of coal enterprises and power generation enterprises must change significantly. It is a big challenge for thermal coal supply chain member enterprises, including coal enterprises, power generation enterprises, and thermal coal transportation enterprises, to make the right decisions in response to the policy influence [29]. Completing the analysis of policy implications will be of great significance in researching the policy mechanism and predicting the developing direction for thermal coal supply chain node enterprises under the policy background. The policy implications are complex and dynamic, so it is difficult for a general model to simulate the changing process of each factor in the thermal coal supply chain under the influence of air pollutant emission reduction policies. Considering the advantages of the SD model on integrity and dynamics during complex analysis, this paper plans to establish a system dynamics model of thermal coal supply chain member enterprises' development processes under the impact of air pollutant emission reduction policies to figure out the policy mechanism and assist decision-making in node enterprises.

A SYSTEM DYNAMICS MODEL OF POLICY IMPACT ON THERMAL COAL SUPPLY CHAIN MEMBER ENTERPRISES

System Dynamics Methodology

SD is a systems modeling and dynamic simulation methodology for analysis of dynamic complexity in socioeconomic and biophysical systems [30]. Based on the principles of system thinking and feedback control theory, system dynamics helps in understanding the time-varying behavior of complex systems [31]. Our SD model is divided into two parts: qualitative analysis and quantitative analysis. The causality diagram is mainly used for the qualitative analysis of the model and the system dynamics flow diagram is used to realize the quantitative analysis. We used a causality diagram to qualitatively analyze the transfer process of impact of air pollution reduction policies on the thermal coal supply chain. The causality diagram achieves a qualitative analysis of complex correlations and influences within various factors of the thermal coal supply-chain system by drawing system factors and the positive or negative causal chains connecting factors. Furthermore, we used a flow diagram to analyze the quantitative relationship between the three variables of the coal subsystem, the power subsystem, and the transportation subsystem. The dynamic flow diagram builds up the quantitative analysis model to simulate and analyze the system's behavior by drawing the visual state variable, the rate variable, and the auxiliary variable, and setting function relationships and initial values among variables through the Vensim_PLE software. The SD model simulated the changing trend of various internal variables under the influence of relevant policies. In addition, we predicted the power structure adjustment and the developing route of coal enterprises by summarizing the simulated results.

The thermal coal supply chain is a system with obvious boundaries, and policy impact on it will be both dynamic and complex. Although other types of dynamic, quantitative modeling can do the impact analysis, an SD model has an advantage in solving dynamic problems because it is the only method that can better reflect forward and reverse policy impact processes along the thermal coal supply chain, and reveal the extent of the influence air pollution reduction policy has on factors in the supply chain system [32]. So this study uses system dynamics to analyze air pollutant emission policies' impact on the development of thermal coal supply chain member enterprises in China, simulating the dynamic transfer process of policy influence. We believe that this SD model can properly analyze the impact of air pollution reduction policies on the development of thermal coal supply chain member enterprises, and simulate the dynamic transfer process of relevant policy influence.

Model Structure Analysis

In order to analyze the transfer process of policy impact within the supply chain system, this paper firstly sorted out possible policy impact types by combining specific action plans and reduction targets mentioned in the above air pollutant emission reduction policies. The air pollutant emission reduction policies are aimed at strengthening the management of air pollution and reducing the concentration of PM10. Coal, as the primary source of atmospheric pollutants, is a key regulatory object of these policies. Policies such as The Action Plan for the Control of Air Pollution will impact the scale, costs, and benefits of coal enterprises, power generation enterprises, and transport enterprises through a variety of means. At present, air pollutant emissions policies such as The Action Plan for the Control of Air Pollution and the action plan implementing rules for the control of air pollution in the Beijing-Tianjin-Hebei area will impact the following aspects of thermal coal supply chain node enterprises:

Industrial Structure Adjustment

The Action Plan for the Control of Air Pollution (or The Action Plan, for short) required the full or partial elimination of small coal-fired boilers in central heating, to be replaced by the use of electricity or natural gas, and forbade new coal boilers in key areas such as Beijing and Tianjin. New projects in Beijing-Tianjin-Hebei, the Yangtze River Delta, and the Pearl River Delta are banned from having supporting self-provided coal-fired power stations. In addition to cogenerations, new thermal-power generation projects are banned. We must speed up the development of hydropower, wind power, solar power, and nuclear power on the basis of safety and efficiency. Nuclear power installed capacity will reach 50 million kilowatts by 2017. Such policies will reduce the capacity of new thermal power generating units and increase the coal enterprise's strategic and operational risk. The Action Plan highlighted how to govern small coal-fired boilers comprehensively, and encouraged the accelerated construction of "coal to gas" and "coal to electricity" projects. Small coal-fired boilers under 10 tons will be eliminated by 2017.

Energy Structure Adjustment

The Action Plan clearly pointed out that the proportion of China's coal consumption will decrease to less than 65% by 2017. China should focus on the orderly development of hydropower, the efficient utilization of geothermal energy, wind energy, solar energy, biomass energy, and the safe development of nuclear power. It must increase the proportion of new power generation and the consumption of renewable energy power by developing projects of wind power and nuclear power generation. Coal should gradually be replaced by increasing both the proportion of outside transmission in the gas supply and the non-fossil energy intensity. The Chinese government recommends the use of electricity and other clean energy instead of coal. Such policies are intended to reduce the proportion of coal consumption in the total energy consumption and limit the relative demand for coal resources.

In addition, The Action Plan stopped imports of inferior foreign coal and restricted sales of bulk coal with high ash and sulfur. These policies would affect the demand-supply condition of coal and the market price of thermal coal, and would bring demand-supply risks to coal enterprises and fuel price risks to power generation companies. The Action Plan also pointed out that by 2017 the proportion of washed coal should rise to 70%, in order to reduce air pollutant emissions in the process of coal combustion. This policy will increase the washing equipment investment and running cost of coal enterprises, thus raising its investment risk.

Technical Renovation

The Action Plan pointed out that coal-fired power generation enterprises should reduce harmful gas emissions by speeding up the desulfuration renovation of thermal power units and formulating strict desulfuration capacity targets. In 2012, the proportion of desulfuration capacity is 56%. This proportion will increase to 70% by 2017. Key areas such as Beijing and Tianjin have to complete the pollution control work of coal-fired power plants before the end of 2015. Meanwhile power generation enterprises should strengthen the development of desulfurization and denitration technology and intensify the exchange of air-pollution control management experience. These policies force coal enterprises to put a lot of money into the desulfurization reform and the development of related emissions reduction technology. These measures will occupy much of the working capital and greatly increase the operational risk, the cash flow risk, and the cost risk for power generation companies.

Market Mechanism Adjustment

The market mechanism adjustment policy includes two aspects of incentives and punishment. In terms of incentives, The Action Plan pointed out that the government will adjust the sales price and perfect the denitration electricity price policy considering the cost of denitration and the local characteristics. Existing thermal power units that adopt new dedusting facilities' renovation technology should be given price supports. In terms of punishment, The Action Plan pointed out that the government should increase the intensity of atmospheric pollutant emission levies and improve atmospheric pollution standards. This kind of market adjustment policy will increase the price risk and cost risk of thermal coal supply chain enterprises.

Vehicle Environmental Management

The Action Plan made it clear that in the future China will strengthen the environmental protection management of vehicles and forbid sales of environmental substandard vehicles. This policy had a limit on coal transportation vehicles. It may affect coal supplies, and even cut down the thermal coal supply chain. The above policies will impact node enterprises of the thermal coal supply chain, and further impact upstream and downstream enterprises driven by principal–agent relationships, benefit distributions, and decision implementations of member enterprises. In the thermal coal supply chain, coal enterprises, power generation enterprises, and coal transportation enterprises are adjoined to one another and jointly realize the processes of thermal coal production, transportation, and consumption. So when one of these three parties is affected by air pollutant emission reduction policies, the other parties will be affected as well.

For thermal power generation enterprises, the policy of shutting down small thermal power plants and the prohibition on new coal-fired power projects will shrink the scale of the thermal power industry, thus reducing coal demand and influencing the income of the coal industry. The reduction in the coal supply will limit power generation enterprises' coal selection. The deviation between the actual burning coal and the design coal can reduce power plant boiler efficiency and improve fuel costs of power generation enterprises. The increasing air pollutant emission pressure will encourage power generation enterprises to conduct the desulfurization and denitration renovation of units. These investments will reduce power generation enterprises' profits. In addition, the rapid rise of clean energy power generation inevitably leads to a decrease in the thermal power installed capacity proportion. This shrinkage will surely decrease the proportion of thermal power generation and the profitability of the thermal power industry. If the thermal power industry pays a fixed proportion of its income towards investment in new construction, with no obvious changes in the unit installed costs of thermal power, the new thermal power installed capacity will be further reduced. For coal enterprises, there are two kinds of air pollution reduction methods: one is coal-production reduction, and the second is coal washing [33]. The increase of coal washing equipment investment will enhance the coal enterprise investment risk, but it can also lower atmospheric pollutant emissions, such as sulfur dioxide and nitrogen oxides, in the process of coal combustion, and cut down the cost risk to thermal power generation enterprises by reducing atmospheric pollutant discharge. Meanwhile, the increase in washed coal will reduce the power supply coal consumption rate, thus reducing the fuel costs of the thermal power industry. Coal washing technology can remove about 50%–80% of the coal ash content. Every 1% reduction in the ash content in thermal coal brings 2–5 grams' reduction in the standard power supply coal consumption rate. With respect to coal supply and demand, the reduction of thermal power generation will cut the coal demand, and thus affect the income and construction investment of the coal industry. The reduction of the coal production growth rate and the import limitation on inferior foreign coal will reduce the growth rate of the coal supply. The coal supply/demand situation tends to loosen, under air pollutant emission reduction policies. The falling coal price will further affect the benefits of coal enterprises.

For coal transportation enterprises, coal-conveying vehicles as important mobile pollution sources will probably be restricted by environmental protection indexes. Vehicles under environmental protection standard could not travel on the road. The policy will block coal transportations and influence the fuel supply of thermal power plants. After analyzing policies, this study uses a causality loop to do the qualitative analysis of air pollution policies' impact on the development of thermal coal supply chain member enterprises in China. The causality loop can be seen in Figure 1. A causality loop is a directed graph applied to analyze the interaction relationships between internal variables of a dynamic system [34]. It consists of several single causality chains. The causality chain representing positive effects is called the positive chain, with "+" next to the arrows. The causality chain representing negative effects is called the negative chain, with "–"next to the arrows. A causality loop is composed of multiple causal chains.

On the basis of the model structure analysis and causality loop, this study divided the system dynamics model of air pollution reduction policies' impact on the development of thermal coal supply chain member enterprises into three modules according to the composition of the thermal coal supply chain. The main factors of the coal module are: amount of washed coal, increasing rate of washed coal, the coal washing cost, proportion of washed coal, proportion goal for washed coal, coal industry profits, coal construction investment, domestic coal production, the coal production rate, net coal import, coal supply, coal demand, coal supply/demand ratio, coal supply/demand ration factor, coal prices, coal consumption proportion of the power industry, and coal consumption of the power supply. The main factors of the power generation module are: thermal-power installed capacity, increase or decrease in the thermal power installed capacity, clean-energy power installed capacity, changing rate of clean-energy power installed capacity, total generated energy, thermal-power generation, proportion of thermal power generation, thermal power profits, thermal power construction investment, grid purchase of thermal power, coal consumption of the power industry, SO_2 emission of thermal power coal consumption, SO_2 emission goal of thermal power coal consumption, desulfuration capacity, desulfuration capacity changing rate, and desulfuration cost. The main factor in the coal transportation module is coal-conveying vehicles' environmental success rate.

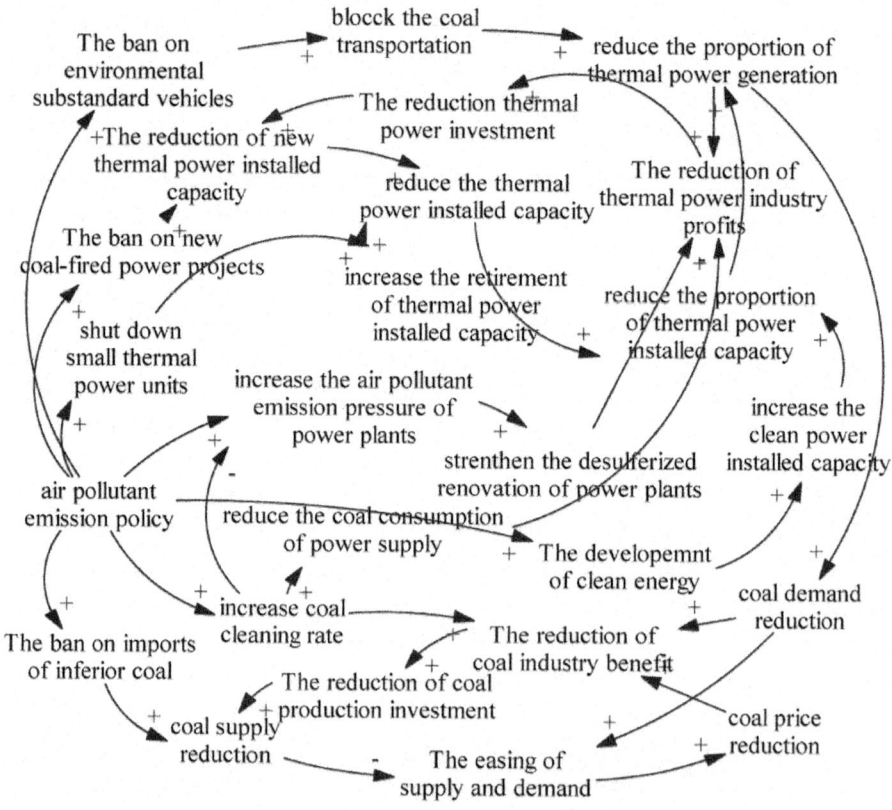

Figure 1. Causality loop of air pollutant emission policies' impact.

The confirmation of model factors and the qualitative analysis using causality loop lays a foundation for further quantitative analysis.

Model Design

After determining the causality loop and the main factors involved in it, this study begins the quantitative analysis of air pollution reduction policies' impact on the node enterprises of the thermal coal supply chain in China by drawing a dynamic flow diagram. A dynamic flow diagram is used to depict the logical relationship between system factors with symbols [35]. This study uses a dynamic flow diagram to clear the feedback form and control the law of the system. Firstly, this paper classified the main factors determined by the causality loop according to their characteristics. Variables representing cumulative results are set as state variables. Variables representing the changing speed rate of state variables are set as rate variables. The rest of the relevant variables are set as auxiliary variables. This study uses Vensim_PLE software to establish the flow graph of air pollutant emission policies' impact on the development of thermal coal supply chain member enterprises. The flow graph is shown in Figure 2.

The above system dynamics model of air pollutant emission policies' impact on the development of thermal coal supply chain member enterprises contains five state variables, six rate variables, and 28 auxiliary variables including time. The upper part of Figure 2 is the coal module of this system dynamics model, and the lower part is the power generation module and the transport module. In Figure 2, the arrow direction indicates the transfer process of air pollutant emission policies' impact between thermal coal supply chain member enterprises. The impact of policies is transmitted through the proportion goal of washed coal, the sulfur dioxide emission of coal-fired power plants, the shutting down capacity of small coal-fired boilers, and the desulfurization grid purchase price influencing the thermal power installed capacity and the coal supply/ demand situation. The Action Plan and other policies have clear plans relative to some key factors such as the goal of air pollutant emission and the installed capacity. So these four factors are expressed as time functions in the quantitative analysis of the system dynamics model. This study used the already built dynamics model of air pollutant emission policies' impact on the development of thermal coal supply chain member enterprises to simulate the transfer process of air pollutant emission policy impact within the thermal coal supply chain during the period 2013–2022 and analyzed changing trends of various factors under the influence of policies during these decades.

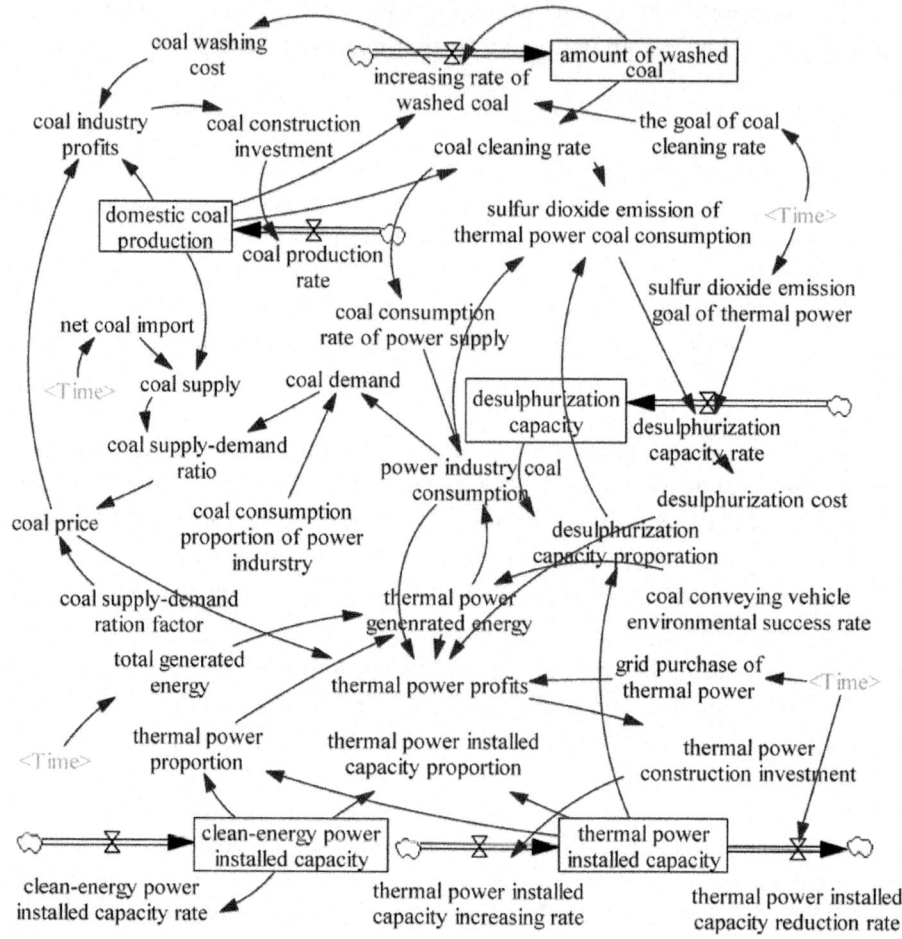

Figure 2. Flow graph of air pollutant emission policies' impact.

In order to perform quantitative analysis on the transmission of atmospheric pollutant emission policies' impact in the thermal coal supply chain and simulate the changing trends of various internal factors under this influence, this study needed to further determine the function of variables on both ends of the arrow in the flow graph. Before determining the functional relationships, the initial value and unit of each factor have to be settled according to statistical data and actual situations. The variable category, variable name, initial value, and unit of variables in this system dynamics model are all shown in Table 1.

Table 1. Variable settings of system dynamics model.

Variable Category	Variable Name	Initial Value	Unit
state variable	amount of washed coal	205,000	10^4 Ton
auxiliary variable	coal washing cost	-	10^8 Yuan
rate variable	increasing rate of washed coal	-	10^4 Ton
auxiliary variable	proportion of washed coal	-	-
auxiliary variable	proportion goal of washed coal	-	-
auxiliary variable	coal industry profits	-	-
auxiliary variable	coal construction investment	-	10^8 Yuan
state variable	domestic coal productions	357,357	10^4 Ton
rate variable	coal production change rate	-	10^4 Ton
auxiliary variable	net coal import volume	-	10^4 Ton
auxiliary variable	coal supply	-	10^4 Ton
auxiliary variable	coal demand	-	10^4 Ton
auxiliary variable	coal supply/demand ratio	-	-
auxiliary variable	coal supply/demand ration factor	-	-
auxiliary variable	coal price	-	Yuan/Ton
auxiliary variable	coal consumption proportion of power industry	-	-
auxiliary variable	coal consumption of power supply	-	g/KWH
state variable	thermal power installed capacity	819,000	MW
rate variable	thermal power installed capacity increasing rate	-	MW
rate variable	thermal power installed capacity reduction rate	-	MW
state variable	clean-energy power installed capacity	323,940	MW
rate variable	clean-energy power installed capacity change rate	-	MW
auxiliary variable	total generated energy	-	10^8 KWh
auxiliary variable	thermal power generated energy	-	10^8 KWh
auxiliary variable	proportion of thermal power generation	-	-
auxiliary variable	proportion of thermal power installed capacity	-	-
auxiliary variable	thermal power generation profits	-	10^8 Yuan

auxiliary variable	thermal power construction investment	-	10^8 Yuan
auxiliary variable	grid purchase of thermal power	-	Yuan/KWH
auxiliary variable	power industry coal consumption	-	10^4 Ton
auxiliary variable	sulfur dioxide emissions of thermal power coal consumption	-	10^4 Ton
auxiliary variable	sulfur dioxide emission goal of thermal power coal consumption	-	10^4 Ton
state variable	desulfurization capacity	753,480	MW
rate variable	desulfurization capacity changing rate	-	MW
auxiliary variable	proportion of desulfurization capacity	-	-
auxiliary variable	desulfurization cost	-	10^8 Yuan
auxiliary variable	coal-conveying vehicle environmental success rate	-	-

In Table 1, the initial value of the amount of washed coal, the dominant coal productions, the thermal power installed capacity, the clean-energy power installed capacity, and the desulfurization capacity are from China's current situation as of 2012. The value of the proportion of washed coal, the sulfur dioxide emission goal of thermal power coal consumption, and the thermal power installed capacity reduction rate are controlled by time according to the relevant requirement of China's air pollutant emission reduction policies. For example, the proportion of washed coal in 2012 is 56%, and The Action Plan pointed out that this proportion will reach 70% by 2017. This model assumes that the proportion of washed coal rises at a constant speed during the period 2012–2017, and by maintaining this rate continues to rise. In the simulation period of 2013–2022, the proportions of washed coal range respectively from 59%, 62%, 65%... 71%... 83% to 86%. For thermal power sulfur dioxide emissions targets, The Action Plan pointed out that the PM10 concentration will drop by more than 10% in 2017 compared to 2012. According to the relevant data, in 2012, China's total carbon dioxide emission was 502 tons, and thermal power's contribution to sulfur dioxide emission accounted for about 50%. This study also assumes that sulfur dioxide emission declines at a constant speed during these five years, and by maintaining this speed continues to decline. So, during the simulation period of 2013–2022, the sulfur dioxide emission goal of thermal power (10,000 tons) ranges from 250, 245, 240... 205 to 200. The Action Plan pointed out that China plans to weed out 50 MW of small thermal power units in 2012–2017. So the rate of thermal power unit elimination for these five years is 10 MW per year. The eliminating speed is expected to accelerate from 2018. So the eliminating capacity of thermal power unit during the period 2018–2022 is 12 MW per year. The total generating capacity and the coal imports during the simulation period are estimated from the regression analysis according to the historical data of 2000–2012. The prediction equations are:

$$G_t = 3218.8 \times (\text{Year} - 1999) + 7613.9 \tag{1}$$

$$I_t = 2.8 + 0.4 \times (\text{Year} - 2012) \tag{2}$$

where Gt is the total generating capacity in year t and It is the coal imports in year t.

The grid purchase of thermal power is estimated according to the desulfurization thermal power electricity price in 2012. So the grid purchase of thermal power is 0.45 yuan per kilowatt-hour in the first five year of simulation, and it decreases to 0.435 yuan per kilowatt-hour after 2017.

The system dynamics model of air pollutant emission reduction policies' impact on the development of thermal coal supply chain member enterprises contains more than 30 functions among relevant variables. Due to the length limitation of this article, we only enumerate those functions with obvious characteristics and great significance to this study. Important functions of impact transfer are introduced as follows:

$$IR = CP \times WG - WC \tag{3}$$

where IR is the increasing rate of washed coal, CP is the domestic coal productions, WG is the proportion goal of washed coal, and WC is the amount of washed coal. Equation (3) shows that the increasing rate of washed coal is the difference between the proportion goal of washed coal and the actual amount of China's washed coal.

$$TS = (1.6 \times TC \times 1.5\%) \times (1 - PW \times 0.3) \times (1 - DS \times 0.9) \tag{4}$$

where TS is the sulfur dioxide emissions of thermal power coal consumption, TC is the thermal power industry coal consumption, PW is the proportion of washed coal, and DS is the desulfurization capacity proportion.

$$SE = 1.6 \times CC \times CS \times (1 - DS \times DE) \tag{5}$$

where SE is the SO_2 emission during the coal combustion of thermal-power industry, CC is the power industry coal consumption, CS is the coal sulfur content (the sulfur content of China's coal is about 1.5%), DS is the desulfuration capacity proportion, and DE is the desulfuration efficiency (China's desulfuration efficiency is about 90%).

$$TP = GP \times TG - CP \times TC / 10000 / 0.7 - DC \tag{6}$$

where TP is the thermal power profits, GP is the grid purchase of thermal power, TG is the thermal power generated energy, CP is the coal price, TC is the thermal power industry coal consumption, and DC is the desulfuration cost. Fuel costs account for 70% of the total costs of thermal power plants.

RESULTS AND DISCUSSION

This study analyzes the air pollution reduction policies' impact on the node enterprises of the thermal coal supply chain by comparing trends of key model factors with and without policy influence. The Vensim_PLE software is used to set up the above system dynamics model and simulate tendencies of the variables during the period 2013–2022. The trend graph of Vensim_PLE software can reflect changing trends more intuitively. This article selects several important factors such as SO_2 emissions of thermal power coal consumption, desulfurization capacity proportion, and the changing rate of thermal power installed capacity to measure the impacts of two typical kinds of air pollution reduction policies, and analyzes the changing developing modes brought by these two policies from multiple perspectives.

The Impact of Sulfur Dioxide Emission Target Policy

The Action Plan pointed out that the PM10 concentration will drop by more than 10% in 2017 compared to 2012. In the preceding part of this study, the SO_2 emission goals of thermal power during the simulation period 2013–2022 have been settled according to the relevant data. The new desulfurization installed capacity of the thermal power industry is fixed to 20,000 MW per year without the influence of this policy. When affected by this emission target policy, the new desulfurization installed capacity of the thermal power industry equals 20,000 (MW per year) multiplied by the ratio of the actual SO_2 emission and the emission goal of the thermal power industry. This paper analyzes the impact of emissions targets policy by comparing the changes of various factors with and without the constraints of this policy. The screenshots of simulation results using Vensim_PLE software are shown in Figure 3. Figure 3 displays the comparison results of SO_2 emissions of thermal power coal consumption (curve 1), the desulfurization capacity proportion (curve 2), and the increasing rate of desulfuration capacity (curve 3). Part (a) shows the simulated results under the impact of SO_2 emissions target policy and the annual SO_2 emission target (curve 4) during the decade simulation period. Part (b) shows the simulated results without policy impact.

Without policy constraints on the thermal power industry, SO_2 emission during coal combustion increases to the highest point of 5.7854 million tons in 2022. The desulfurization capacity proportion remains at around 89% with no obvious changes. The new desulfurization capacity increases from 17,566 in 2012 to 32,227 MW in 2022. Under the constraint of pollution emission targets, the SO_2 emission reduced greatly in 2013–2017 and then started to level off. It decreased to the lowest point of 2.8284 million tons in 2019. The desulfurization capacity proportion was greatly increased in 2013–2017. This proportion will reach 99% since 2017. There is no obvious difference between the two simulation results of thermal power installed capacity increasing rate with and without the constraint of pollution emission policies.

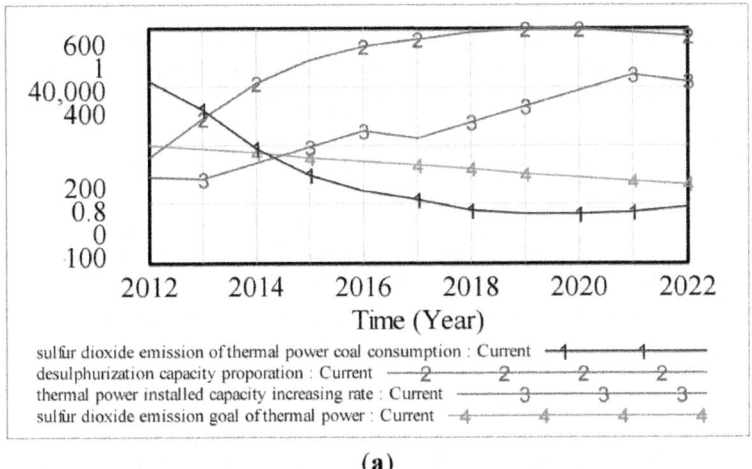

(a)

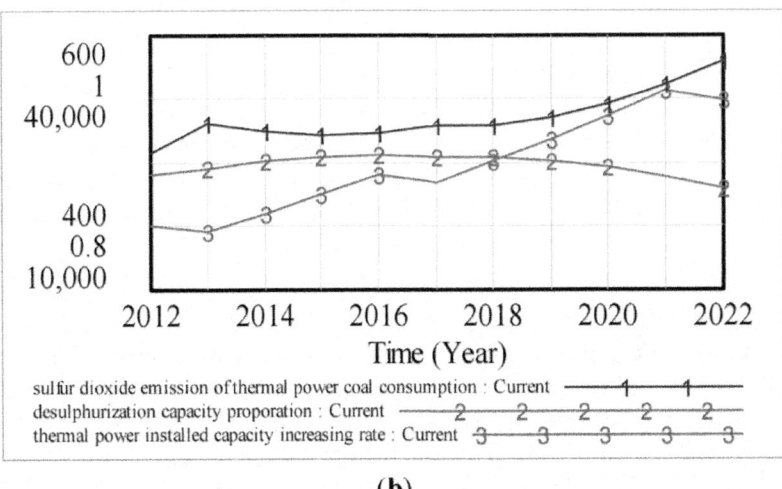

(b)

Figure 3. (a) Simulated results under the influence of SO_2 emissions target policy; (b) Simulated results without the influence of SO_2 emissions target policy.

We can tell from the compared results that SO$_2$ emission target policies will play an important role in reducing air pollutant emissions. SO$_2$ emission during coal combustion of the thermal power industry will be reduced by about 3 million tons per year. This emission target policy also can lead to a significant increase in the desulfurization capacity proportion. It is estimated that China will nearly complete the desulfurized renovation of the thermal power industry in 2017 under the influence of SO$_2$ emission target policy. As the result of space limitations, this paper only contrasts the changes in key factors. The SO$_2$ emission target policy can also decrease the proportion of thermal power installed capacity by reducing coal industry profits and thermal power construction investment. In 2012, China's thermal power installed capacity proportion is 71.6%. According to the simulated result, this proportion will decline steadily during the period 2013–2022. It will decrease to the lowest point of 62% in 2022. These structural changes are driven by the need for air pollution mitigation, not only in China but also in Europe and many developed countries [36]. After adjusting the parameters and variables based on the actual situation, this SD model can be used to simulate the policy impacts in all these countries. With the increasing of energy conservation and emission reduction policy's intensity, countries like China will implement sustainable development strategies and accelerate the transformation of the power industry.

The Impact of Coal Washing Proportion Target Policy

The Action Plan clearly set up the target that washed coal should account for more than 70% of total coal production in 2017. In the preceding part of this study, the coal washing proportion goals of thermal power during the simulation period of 2013–2022 have been settled according to the relevant data. This paper assumes that the coal washing proportion target will increase at a constant speed from 56% in 2012 to 71% in 2017 and keep increasing at the same speed during the rest of the simulated period under the impaction of the coal washing proportion target policy. Without the policy's influence, the coal washing proportion target will be fixed at 56% during the whole simulated period. The coal washing proportion goal influences the actual amount of washed coal by determining the increasing rate of washed coal. The increasing rate of washed coal equals the domestic coal production multiplied by the coal washing proportion goal. The amount of washed coal is the integral of washed coal's increasing rate. This paper analyzes the impact of proportion targets policy by comparing the changes of various factors with and without the constraints of this policy. The screenshots of simulation results using Vensim_ PLE software are shown in Figure 4. Figure 4 displays the comparison results of actual coal washing proportion (curve 1), SO$_2$ emission during the coal combustion of the thermal power industry (curve 2) and the increasing rate of thermal power installed capacity (curve 3). Part (a) shows the simulated results under the impact of coal washing target policy and the annual goal of coal washing proportion (curve 4) during the simulation period. Part (b) shows the simulated results without the policy's impact.

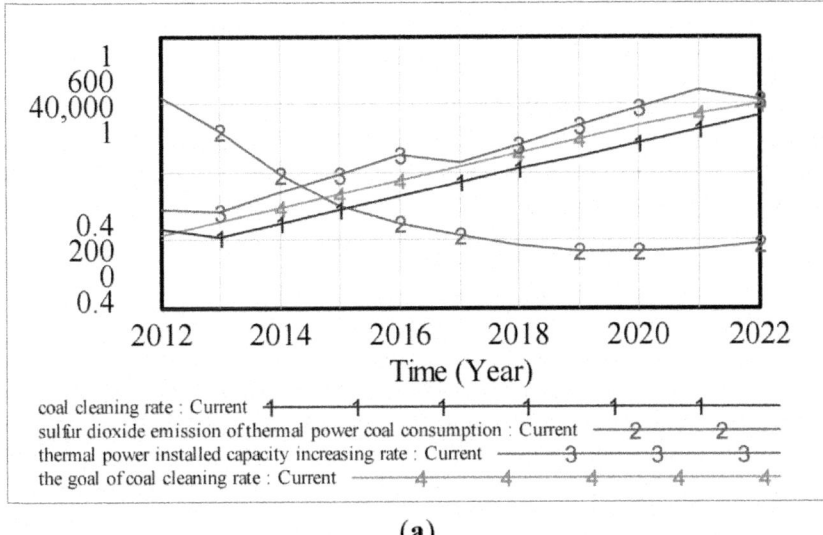

(a)

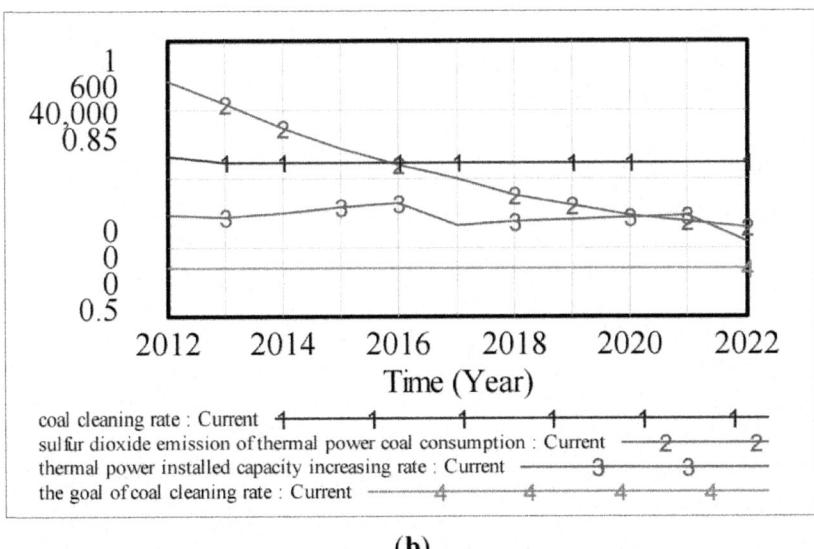

(b)

Figure 4. (a) Simulated results under the impact of washing coal proportion target policy; **(b)** Simulated results without the impact of washing coal proportion target policy.

Without the constraints of coal washing proportion target policy, the actual coal washing proportion is almost running at 56% during the simulated period. SO_2 emission from the coal combustion of the thermal power industry has a slow decline of about 2 million tons. In addition, the new thermal power installed capacity is fluctuating around 14,500 MW for the simulated period. Under the constraint of the coal washing proportion target, the actual proportion of washed coal increases at a constant speed from 56% in 2012 to 82% in 2022. SO_2 emission from the coal combustion of the thermal power industry has a sharp decline of more than 2 million tons and eventually stabilizes at about 2.9 million tons during the period 2013–2022. The new thermal power installed capacity will increase from 14,479 MW in 2012 to 30,720 MW in 2022.

We can tell from the simulated results shown in Figure 4 that the proportion target of washed coal established by air pollution reduction policies plays an important part in reducing SO_2 emissions. SO_2 emissions from the coal combustion of the thermal power industry will be reduced by about 2 million tons per year, which is less than the emission reduction driven by the emission target policy mentioned above. The formulation of a coal washing proportion goal makes the actual amount of washed coal increase by nearly 30%. In addition, the increase in washed coal can reduce the coal consumption rate in the power supply and improve the profits of the thermal power industry, and eventually lead to a slight rise in new thermal power installed capacity. At the same time, the sharp rise in clean energy power capacity will ensure that the power structure can shift in the sustainable development direction. The coal washing proportion goal policy can reduce the thermal power enterprise's fuel costs and improve its profits so that the thermal power industry can put more money into the desulfurization and denitration reform of coal-fired units. In general, a coal washing proportion goal policy can promote energy conservation and the emissions reduction technologies of the coal and thermal power industries by economic means. More applications of energy conservation and emission reduction technology can reduce the energy intensity of the industry and reduce air pollutant emissions fundamentally.

CONCLUSIONS

This paper simulated the air pollution reduction policies' impact on the thermal coal supply chain members in China by establishing SD models. These policies will greatly impact the development patterns of coal enterprises, power enterprises, and coal transportation enterprises. Moreover, the influence will transmit to the upstream and downstream enterprises along the thermal coal supply chain driven by business transactions and decision implementation. Besides China, many other countries (such as Japan and EU-27) have also made air pollution control policies, just like The Action Plan, to set emission targets and restrict coal consumption [37]. According to the simulated results, air pollution reduction policies can significantly improve air quality by promoting power structure adjustment and improving energy efficiency. As the main way of developing and utilizing non-fossil energy, the power industry will take the clean development route to coordinate sustainable development strategy. Under the pressure of air pollution reduction, the thermal power industry will implement the transformation of energy-saving and emission reduction ahead of time. These policies also provide coal washing technology and new energy power generation with good development opportunities. Renewable energy and nuclear electricity generation will have to a develop quickly to accelerate the structure adjustment of the electric power industry and realize the transformation of energy from fossil fuels towards clean energy. Therefore our simulation analysis of different policy interventions has meaning for countries that have not yet established their own air pollution control policies (such as India). In the future, in the promotion of long-distance transmission of electricity and coal, the distribution of coal in the power industry will be further optimized and the development of thermal coal supply chain members will be affected constantly. We can continue to analyze their development path under the new policy environment by adjusting the parameters and variables of the existing SD model.

ACKNOWLEDGMENTS

Project supported by the Fundamental Research Funds for the Central Universities of China (No. MS201439).

AUTHOR CONTRIBUTIONS

Xiaopeng Guo designed the study and revised the manuscript. Xiaodan Guo participated in designing the study, interpreted the data, wrote the manuscript and revised it until its final version. Jiahai Yuan provided good advices for conclusions and revised the manuscript.

REFERENCES

1. Xue, B.; Geng, Y.; Katrin, M.; Lu, C.; Ren, W. Understanding the Causality between Carbon Dioxide Emission, Fossil Energy Consumption and Economic Growth in Developed Countries: An Empirical Study. *Sustainability* **2014**, *6*, 1037–1045.

2. Frances, C.M. Climate Change and Air Pollution: Exploring the Synergies

and Potential for Mitigation in Industrializing Countries. *Sustainability* **2009**, *1*, 43–54.

3. Zhao, X.; Ma, Q.; Yang, R. Factors influencing CO_2 emissions in China's power industry: Co-integration analysis.*Energy Policy* **2013**, *57*, 89–98.

4. Zhao, X.; Thomas, P.L.; Cui, S. Lurching towards markets for power: China's electricity policy 1985–2007. *Appl. Energy***2012**, *94*, 148–155.

5. Peggy, M.; Kenneth, B.K. Modelling tools to evaluate China's future energy system—A review of the Chinese perspective. *Energy* **2014**, *69*, 132–143.

6. Liu, L.; Zong, H.; Zhao, E.; Chen, C.; Wang, J. Can China realize its carbon emission reduction goal in 2020: From the perspective of thermal power development. *Appl. Energy* **2014**, *124*, 199–212.

7. Tan, Z.; Zhang, H.; Shi, Q.; Xu, J. Joint optimization model of generation side and user side based on energy-saving policy. *Electr. Power Energy Syst.* **2014**, *57*, 135–140.

8. Cai, L.; Guo, J.; Zhu, L. China's Future Power Structure Analysis Based on LEAP. *Energy Sources Part A Recovery Util. Environ. Eff.* **2013**, *35*, 2113–2122.

9. Zheng, M.; Zhang, K.; Dong, J. Overall review of China's wind power industry: Status quo, existing problems and perspective for future development. *Renew. Sustain. Energy Rev.* **2013**, *24*, 379–386.

10. Hirota, K. Comparative Studies on Vehicle Related Policies for Air Pollution Reduction in Ten Asian Countries.*Sustainability* **2010**, *2*, 145–162.

11. Cai, W.; Wang, C.; Zhang, Y.; Chen, J. Scenario analysis on CO_2 emissions reduction potential in China's electricity sector. *Energy Policy* **2007**, *35*, 6356–6445.

12. Mathews, A.J.; Tan, H. The transformation of the electric power sector in China. *Energy Policy* **2013**, *52*, 170–180.

13. Wen, Z.; Li, H. Analysis of potential energy conservation and CO_2 emissionsreductionin China's non-ferrous metals industry from a technology perspective. *Int. J. Greenh. Gas Control* **2014**, *28*, 45–56.

14. Wang, K.; Wang, C.; Lu, X.; Chen, J. Scenario analysis on CO_2 emissions reduction potential in China's iron and steel industry. *Energy Policy* **2007**, *35*, 2320–2335.

15. SheikhiFini, A.; Parsa Moghaddam, M.; Sheikh-El-Eslami, M.K. A dynamic model for distributed energy resource expansion planning

considering multi-resource support schemes. *Electr. Power Energy Syst.* **2014**, *60*, 357–366.

16. Zhu, H.; Huang, G. Dynamic stochastic fractional programming for sustainable management of electric power systems. *Electr. Power Energy Syst.* **2013**, *53*, 553–563.

17. Salman, A.; bin Razman, M.T. Using system dynamics to evaluate renewable electricity development in Malaysia.*Renew. Electr. Dev.* **2013**, *43*, 24–39.

18. Li, L.; Sun, Z. Dynamic Energy Control for Energy Efficiency Improvement of Sustainable Manufacturing Systems Using Markov Decision Process. *Cybern. Syst.* **2013**, *43*, 1195–1205.

19. Garcia, E.; Mohanty, A.; Lin, W.; Cherry, S. Dynamic analysis of hybrid energy systems under flexible operation and variable renewable generation-Part II: Dynamic cost analysis. *Energy* **2013**, *52*, 17–26.

20. Santiago, M.; Luis, J.M.; Felipe, B. A system dynamics approach for the photovoltaic energy market in Spain. *Energy Policy* **2013**, *60*, 142–154.

21. Ali, K.; Mustafa, H. Exploring the options for carbon dioxide mitigation in Turkish electricpower industry: System dynamics approach. *Energy Policy* **2013**, *60*, 675–686.

22. Feng, Y.; Chen, S.; Zhang, L. System dynamics modeling for urban energy consumption and CO_2 emissions: A case study of Beijing, China. *Ecol. Modell.* **2013**, *252*, 44–52.

23. Li, F.; Dong, S.; Li, Z.; Li, Y.; Wan, Y. The improvement of CO_2 emission reduction policies based on system dynamics method in traditional industrial region with large CO_2 emission. *Energy Policy* **2012**, *51*, 683–695.

24. Nastaran, A.; Abbas, S. A system dynamics model for analyzing energy consumption and CO_2 emission in Iranian cement industry under various production and export scenarios. *Energy Policy* **2013**, *58*, 75–89.

25. Özer, B.; Görgün, E.; Incecik, S. The scenario analysis on CO_2 emission mitigation potential in the Turkish electricity sector: 2006–2030. *Energy* **2013**, *49*, 395–403.

26. Frederick, A.A.; David, O.Y.; Alex, A.P. A Systems Dynamics Approach to Explore Traffic Congestion and Air Pollution Link in the City of Accra, Ghana. *Sustainability* **2010**, *2*, 252–265.

27. Lin, B.Q.; Moubarak, M. Renewable energy consumption—Economic growth nexus for China. *Renew. Sustain. Energy Rev.* **2014**, *40*, 111–117.

28. Bloch, H.; Rafiq, S.; Salim, R. Economic growth with coal, oil and renewable energy consumption in China: Prospects for fuel substitution. *Econ. Modell.* **2015**, *44*, 104–115.

29. Shen, J.F.; Xue, S.; Zeng, M.; Wang, Y.; Wang, Y.J.; Liu, X.L.; Wang, Z.J. Low-carbon development strategies for the top five power generation groups during China's 12th Five-Year Plan period. *Renew. Sustain. Energy Rev.* **2014**, *34*, 350–360.

30. Weller, F.; Cecchini, L.A.; Shannon, L.; Sherley, R.B.; Robert, J.M.; Altwegg, R.; Scott, L.; Stewart, T.; Jarre, A. A system dynamics approach to modelling multiple drivers of the African penguin population on Robben Island, South Africa.*Ecol. Modell.* **2014**, *277*, 38–56.

31. Jose, B.C.; Tan, R.R.; Culaba, A.B.; Ballacillo, J.A. A dynamic input–output model for nascent bioenergy supply chains.*Appl. Energy* **2009**, (Suppl. 1), S86–S94.

32. Haghshenas, H.; Vaziri, M.; Gholamialam, A. Evaluation of sustainable policy in urban transportation using system dynamics and world cities data: A case study in Isfahan. *Cities* **2014**. in press.

33. Mao, X.Q.; Zeng, A.; Hu, T.; Xing, Y.K.; Zhou, J.; Liu, Z.Y. Co-control of local air pollutants and CO_2 from the Chinese coal-fired power industry. *J. Clean. Prod.* **2014**, *67*, 220–227.

34. Wang, S.; Xu, L.; Yang, F.L.; Wang, H. Assessment of water ecological carrying capacity under the two policies in Tieling City on the basis of the integrated system dynamics model. *Sci. Total Environ.* **2014**, *472*, 1070–1081.

35. Rehan, R.; Knight, M.A.; Unger, A.J.A.; Haas, C.T. Financially sustainable management strategies for urban wastewater collection infrastructure–development of a system dynamics model. *Tunnell. Undergr. Space Technol.* **2014**,*39*, 116–129.

36. Bollen, J.; Brink, C. Air pollution policy in Europe: Quantifying the interaction with greenhouse gases and climate change policies. *Energy Econ.* **2014**, *46*, 202–215.

37. Kanada, M.; Fujita, T.; Fujii, M.; Ohnishi, S. The long-term impacts of air pollution control policy: Historical links between municipal actions and industrial energy efficiency in Kawasaki City, Japan. *J. Clean. Prod.* **2013**, *58*, 92–101.

Chapter 4

SUSTAINABLE LIVING IN AFRICA: CASE OF WATER, SANITATION, AIR POLLUTION AND ENERGY

David O. Omole[1,2] and Julius M. Ndambuki[3]

[1]Department of Civil Engineering, Tshwane University of Technology, Private Bag X680, Pretoria 0001, South Africa

[2]Department of Civil Engineering, Covenant University, P.M.B. 1023, Ota, Ogun State +234, Nigeria

[3]Department of Civil Engineering, Tshwane University of Technology, Private Bag X680, Pretoria 0001, South Africa

ABSTRACT

The study reviewed developmental challenges confronting African countries with specific reference to the availability of potable water, sanitation, energy, water and ambient air. It showed the conflict between the need to exploit environmental capital in order to keep up with the pace of human development activities and the need to utilize resources sustainably. Hitherto, the cost of this development has been at the expense of public health and cleaner environment. The outcome demonstrates the need for a change of approach in the way and manner that environmental resources are exploited for developmental purposes. Two concepts for addressing these problems were discussed. These are the "soft path" approach and the trialog model. The former places high priority on the proper use and management of existing infrastructure or resources rather than acquisition or exploitation of more infrastructure or resources. The latter concept addresses the principle of resource governance through the application of an understanding of the complex relationship between the main stakeholders—government, science, and society. Case studies on the practicality of these concepts were also highlighted and discussed.

INTRODUCTION

Sustainable development has been classified into three broad areas–environmental, economic and socio-political. The concept of sustainable development itself discusses ways of meeting current human needs without depriving future generations of the right and access to the same resources [1,2,3]. In the attempt to satisfy current human needs, a great deal of environmental capital (naturally occurring resources) is used up, sometimes unsustainably, for the provision of infrastructure such as potable water supply, energy supply, and transportation. With constant increase in human population and rural-urban migration, the demand for environmental capital is on the rise [4,5]. Presently, an estimated half the world's population lives in urban areas [5,6]. The needs of these people for water, energy, transportation, and all that makes for life's comfort, constantly puts pressure on environmental capital. The competition for environmental capital is further illustrated when Africa's current population of slightly over a billion people (about 14% of the global population) is compared with the projected population of 1.8 billion people (or 20% of the projected global population) by 2050 [7,8,9]. This projected population explosion in Africa is attributable to the fact that Africa currently has the highest fertility (4.9 births per woman) and growth rates (2.2%) in the world [9]. Unfortunately, the geometric increase in the African population does not correspond with economic growth - a disparity which leads to unsustainable competition for environmental capital [10] (pp. 1563–1564). Using 2009 gross domestic product (GDP) estimates, 70% of all African countries had an average GDP of $1408 (purchasing power parity). The poverty illustrated by this figure becomes accentuated when contrasted with the highest GDP value of $91,379 per capita in the world from Qatar or the per capita GDPs of China and India with values of $6828 and $3296 despite high populations figures of 1.35 and 1.24 billion people respectively [10,11]. This relatively high level of economic poverty in Africa is reflected in every other aspect of human life, such as education, health, and the provision of food and shelter. Research has shown that public health is intricately linked to housing and related infrastructural facilities, such as water, sanitation, and energy [12,13,14], thus, highlighting the need to provide sustainable solutions to these problems. This study, therefore, assesses the ways through which basic human needs, such as water, sanitation, and energy, are currently being met in Africa, *vis-à-vis* the adverse effect that such practices have on humans and on the environment. Alternative approaches to tackling these problems are also discussed.

ENVIRONMENTAL CHALLENGES IN AFRICA

Access to Potable Water and Sanitation

Water-related diseases have been identified as the leading cause of human mortality in the world [15,16]. About half of all hospitalized people in Africa were reported to be suffering from water or sanitation-related illnesses [15,17]. Challenges of access to potable water supply are more pronounced in rural areas, where women spend more than a quarter of the productive hours of each day on fetching water from sources, which may not be hygienic [18,19]. This is in contrast to the United Nations standard, which specified that each household should have water delivered to the occupants via *in situ* taps or via a water source that is within 100 meters or 5-minute total collection time [18,19,20,21,22].

An analysis of available data establishing the link between the level of water/sanitation crises and the level of human development in three selected African countries is shown in Figure 1. These countries are Tunisia, Swaziland, and Niger, representing the high human development index (HHDI), the medium human development index (MHDI) and the low human development index (LHDI) respectively [11,23,24]. The selection was based on the country having the poorest record in each category. While the population without access to water and sanitation in Tunisia is very low, it can be observed that the population of persons with problems of water and sanitation increased as the level of HDI decreased (Figure 1). The worst record was in the Niger Republic where 64.1% and 89.3% of its population lacked access to clean water and improved sanitation respectively.

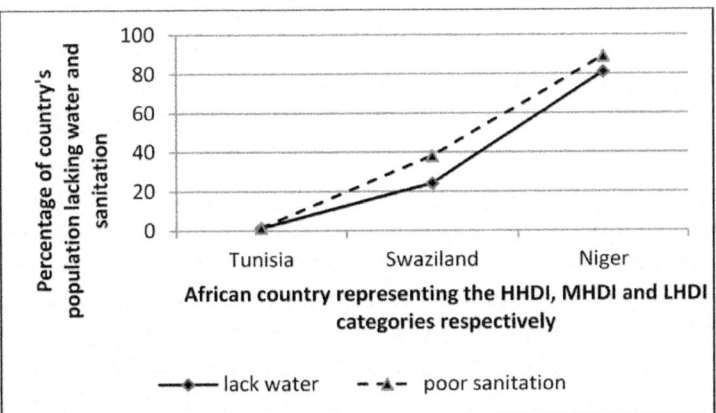

Figure 1. Percentage population of selected country within each HDI category that lack access to clean water and sanitation [Source: [11] (pp. 143–145)].

Other African countries having acute water scarcity and sanitation records include Rwanda, Uganda, Democratic Republic of Congo (DRC), Ethiopia, Central African Republic, Angola, Burundi, Sierra Leone, and Zambia [11,25]. These are countries with more than half of the citizens living without sufficient water [11,26,27]. Some of the identified causes of water scarcity and sanitation problems in these African countries include:

- *Pollution and depletion of available resources through human activities*: about 80% of untreated domestic wastewater in Africa is released into surface water bodies, thus depleting available freshwater [28,29]. This is besides wastewater from agricultural and industrial activities [28] (pp. 36–38).

- *Insufficient finances*: development of water resources is capital intensive. It requires huge investments and subsequent maintenance efforts. Most African countries cannot afford the required investments, thus, limiting their water developmental projects and distribution capacities [29,30,31].

- *Water losses*: infrastructural facilities, such as pipelines are often vandalised by people seeking illegal connections to public water services. Such activities lead to water wastages and financial losses [26,27,32,33].

- *Weak water governing institutions*: some of the water corporations in African countries are operationally limited by factors such as lack of data, inept personnel, energy shortages, and limited finances [25,32,34,35].

Environmental Pollution and Public Health

Anthropogenic interferences such as agricultural, industrial, transportation and domestic activities have been instrumental to environmental pollution [36,37,38]. Of particular interest in this section is pollution affecting the atmosphere. Common health challenges arising from air pollution include pneumonia, tuberculosis, cataracts, upper air-way cancer and asthma, especially in children [39,40]. Air pollution can be classified into two categories, namely outdoor air pollution (OAP) and indoor air pollution (IAP). Air pollution arising from agricultural, industrial and transportation activities are outdoor pollutions. Studies have shown that OAP arising from vehicular emissions, which often contain carbon monoxide, sulphur dioxide, nitrous oxide, volatile organic compounds, and lead, are the highest sources of OAP [36,41,42,43]. Much of these emissions contribute to global warming. In addition, air pollution arising from domestic activities poses another set of challenges because they occur in relative isolation in the confines of individual households. Globally, an estimated 1.6 million people die annually from IAP, caused mainly by fumes generated by cooking activities or through smoke that is introduced into living quarters for the purpose of repelling mosquitoes [39,40,44]. About a quarter of global mortalities arising from IAP reportedly occur in Sub-Sahara Africa [40,44,45]. Analysis of available data on the link between the mortalities arising from air pollution and the level of human development among African countries is shown inFigure 2 [11] (pp. 150–153). Tunisia, Ghana and Niger had the highest mortalities arising from IAP in the HHDI, MHDI, and LHDI categories respectively, while Libya, Egypt, and Djibouti had the highest mortalities arising from OAP in the HHDI, MHDI, and LHDI categories respectively [11,45].

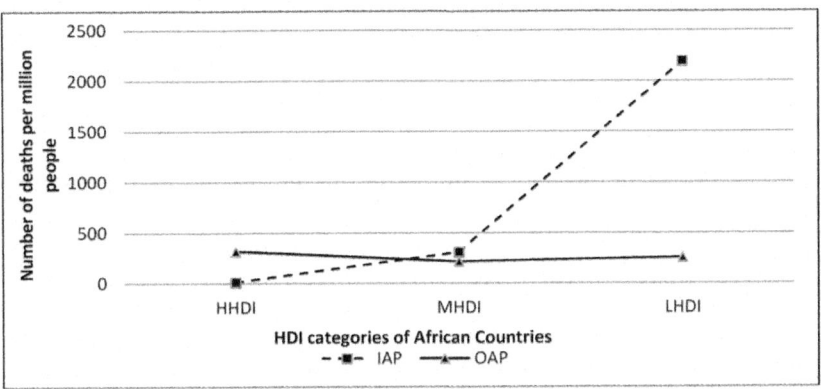

Figure 2. Deaths due to indoor and outdoor air pollution in 2004 [Source: [11] (pp. 150–153)].

Although the total volume of OAP generated per annum exceeded IAP [36], more people reportedly died from IAP than OAP in the MHDI and LHDI categories. The higher mortality rate from IAP could be attributed to poor ventilation, because polluted air is not easily diffused in confined spaces. This was not the case, however, in the HHDI category (Figure 2) as there were more reported cases of death from OAP than IAP. Generally there were higher mortalities due to air pollution in the LHDI countries than the HHDI and MHDI categories, with the worst cases happening in Sierra Leone, Niger, Angola, Burundi, Guinea Bissau, Liberia, Democratic Republic of Congo, Rwanda, Mali, and Burkina Faso, respectively [11,46].

Sustainable Energy Development

There is a positive correlation between energy (generation and consumption) and human development as demonstrated by the fact that developed countries in North America (USA and Canada), European countries, and China, generated 24.6%, 27.5% and 17.2% respectively of the world's total installed energy capacity [47,48]. Comparatively, all the African countries generated just 2.66% of global energy produced in 2009 [48,49]. Energy generation in Africa is broadly classified into two types, based on the source. These sources are fossil fuel (oil, gas, and coal) and biomass. These energy sources are also geographically distributed. While fossil fuels are the predominant fuel sources in Northern and Southern Africa, Sub-Saharan Africa, where over 70% of the African population is concentrated, predominantly utilizes biomass [49,50,51]. Although fossil fuels are known to impact negatively on the environment, they are still the predominant global fuel sources as they constituted 56.5% of the total fuel consumed in the world as at 2009 [47,52,53]. Furthermore, methods of biomass usage in sub-Sahara Africa, such as open air burning, are detrimental to the environment [51]. Generally, solid fuel is the fuel of choice among most poor Sub-Sahara African people because it is cheap to procure and easy to use. However, solid fuels have the highest IAP potentials. A comparison of the choice of fuel by the population range of persons that use them is indicated in Figure 3. A further indication of income level and the choice of fuels are also shown in Figure 3.

It is estimated that three billion people used solid fuels as at 2005, and over 500 million of them resided in Africa [40,49,53]. This practice contributes directly to deforestation and consequently, climate change. It is also a cause of respiratory health challenges among its users.

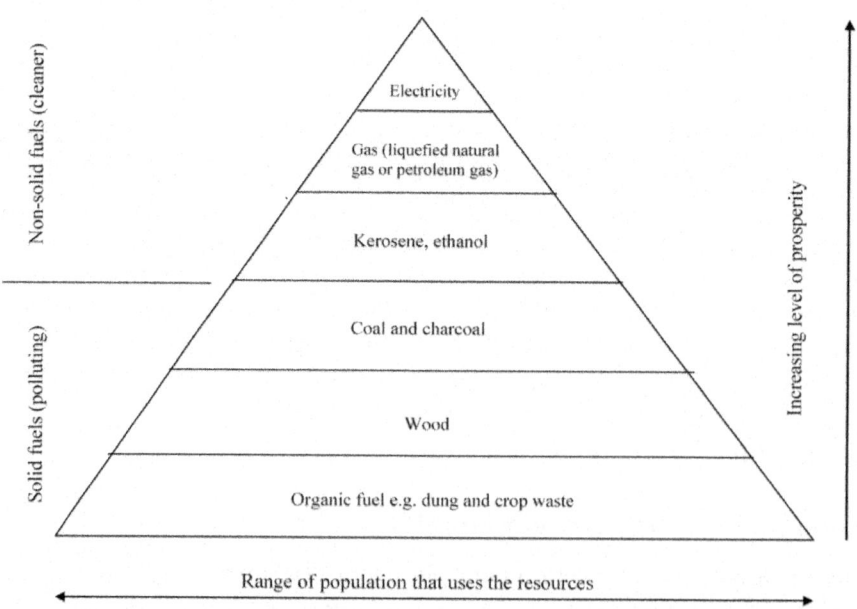

Figure 3. Relationship between fuel sources and income level (Source: [40]).

The energy sources indicated in the upper rung of the chart (Figure 3) are relatively cleaner but more expensive. At the apex of the chart is electricity, which is used by a relatively lower number of people although it is a relatively cleaner source of energy. Electricity, especially when generated by wind, solar or biogas energy, has lesser adverse effect on the environment. The total installed capacity for electricity generation in Africa as at 2009 was 3% of the global total. This has led to Africa becoming the continent with the highest energy intensity in the world [47,49]. This means that due to shortages in clean energy production, Africa's energy demand is being met in unsustainable ways such as biomass burning. Although Africa contributes nearly 16% of global fossil fuel production, neither the resource, nor the revenues from its sales has been translated into equivalent development for the continent [47,49]. Much of the revenue is lost due to misappropriation and mismanagement practices, which has further impoverished the African people [35,40,49,54,55,56].

DISCUSSION

The Way Forward

The continued existence of man is dependent on the sustenance and manner of utilization of important natural resources. Discharging industrial and domestic wastewater into fresh water bodies, cutting down trees for biomass burning and depletion of the ozone layer with greenhouse gases are examples of the misuse of natural resources. This paradox therefore presents a situation where there is conflict in the need to use and also protect and conserve environmental capital. The solution to this problem could be found in a couple of concepts proposed by researchers. One of such concepts is the "soft path" approach to resource management. Originally, the concept of the "soft path" advocated the optimum management and proper use of water in contrast to the traditional "hard path" that advocated more supply of water through investment in physical infrastructure [30]. Soft path approach has been found to be very effective and applicable to other aspects of human endeavors. More results are being achieved at lesser cost since the paradigm shift is aimed at adjusting certain variables in what is presently available, rather than exploiting more of nature's resources [31] (p. 1).

Another concept, which is quite similar to the "soft path" approach, is the trialog model for resource governance. The latter was proposed by the Council for Scientific and Industrial Research (CSIR) in South Africa and it has received positive attention among applied and social scientists [57,58,59,60]. The model proposes a democratic means for determining how natural resources should be used for the common good by identifying the three stakeholders and the interface that exists between them (Figure 4).

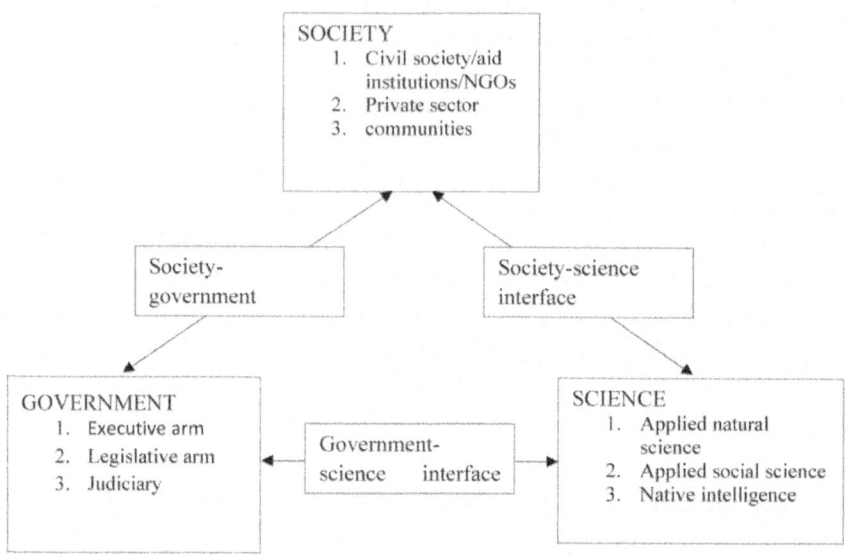

Figure 4. The trialog model [Source: [60] (p. 376)].

The trialog model is not farfetched from the better known public-private-partnership (PPP) concept which emphasizes the relationship between government and non-governmental organizations [55,61]. The main difference between PPP and trialog model, however, is that the trialog model emphasizes the role and importance of a third party (science) in the partnership [58,59]. Trialog model also emphasizes the importance of public perception as well as the promotion of democratic principles among the three parties [60] (pp. 375–377). Thus, trialog model is a simplified representation of the complex relationships between government, society and science. It advocates that government (as embodied by the executive, legislative, judicial, administrative, regulatory, and law enforcement organs) cannot effectively govern natural resources without scientific input or societal consent and cooperation. For instance, societal perception is a strong factor in the implementation of any policy. No matter how rational the scientific recommendation, society perception may make its implementation difficult or even impossible. In addition, organized science (natural and social) form the basis for informed decision taking. However, informal science or native intelligence has its usefulness, especially in the area of data gathering. Society has always been the custodian of natural resources and could offer its opinion when sought. This opinion when acted upon by organized science becomes a useful piece of information. The trialog model, though initially applied to water resource governance in South Africa, is also very relevant to governance issues concerning other natural resources such as ambient air and energy sources.

Global Case Studies

Some case studies that illustrate the application of both the trialog model and the soft path approach are found in different parts of the world. Conley and Dukkipati [61] (pp. 4–11) provide insight into the differing levels of cooperation between different European governments and non-governmental organizations (NGOs), who have a common goal of generating positive impact on the lives of ordinary people in different parts of the world. One such cooperation includes the collaboration of German and Netherland governments with the Bill and Melinda Gates foundation in solving health and environment related problems in developing countries through research as well as actionable programs. While the German government pledged €14 billion for the program, the Bill and Melinda Gates foundation pledged to match Germany's contribution towards the same program, thereby bringing the total budget for the program to €28 billion. These funds were disbursed to research based institutions, as well as indigenous non-governmental organizations in targeted countries for implementation. Regardless of the good intentions however, records indicate that there has been cases of failure of the program in some locations due to the singular fact that the members of the society who are meant to benefit from such people oriented programs distrusted the gestures and intents of the sponsors [62,63]. In Northern Nigeria, it was reported that such program was boycotted en-masse due to religious and cultural reasons [63] (p. 9). In Spain, it was also reported that the program which was developed to effectively manage the groundwater and ecosystem was met with conflicts because the stake holders were not carried along in the decision-making and planning phases of the program [58] (pp. 3389–3396). These problems could be avoided if social scientists/researchers could advise the sponsors on how best to implement such programs by considering social factors. Additionally, the government and civil societies could help to allay public fear through massive public campaigns and enlightenment programs. These examples illustrate the need for application of democratic principles, as espoused in the trialog model, in resource management and implementation. Conversely, some successes have been recorded in initiatives that involve the partnership between governments, NGOs, the society at large and researchers. It is reported that the highest successes in this regard is found in Netherlands and United Kingdom where the health, education, justice, transport, utilities, social housing and defense sectors of the economy have benefitted from public-private-society cooperation [55,64]. In Netherlands, the construction of new public wastewater management systems valued at €1.58 billion was conceded to a private firm for a period of 30 years by the government [55,65]. Concession means that government allows private firms to build, operate and transfer public utilities and/or services over a fixed period of time following which such facilities are handed back to the government. During the concession period, government retains ownership of such facilities or projects but its management and profit goes to the private organization during the concession period. In this arrangement, the government gets to provide better infrastructure and employment opportunities to its constituents, while saving public funds that could have been expended on such projects. The governments also serve as industry regulators and mediators between the private firms and society. The private firms make profit as a direct result of implementing cutting edge research/scientific innovations while the constituents/society get better services and an improved environment. In this sort of situation however, there is always the need for public enlightenment because the public usually has distrust for such programs due to the fact that concessions are often confused with privatizations [55] (p. 401). In the energy sector, a private German firm known as EnviTec in collaboration with the India's Malavalli Power Plant Private Limited (MPPPL) has helped to provide electricity for over 180,000 Indian households using decentralized biogas technology [61] (p. 8).

Soft path approach to resource management is illustrated when efficient management of resources is favored above abstraction or exploration of more resources. As an example, researchers have found ways to reduce fossil fuel and water consumption through the use of biogas systems in Africa. Biogas Technologies West Africa Limited (BTWAL) utilized human wastes received in specially constructed toilets to generate biogas. The generated biogas is in turn used for electricity generation that provides lighting and drives the mechanical pumps used to supply water to the toilets [66,67]. In addition, nearly all the water used in the toilets are recovered and re-used within the toilets, thereby saving a lot of valuable fresh water. This technology has been implemented in schools, hospitals and military/Para-military barracks, in Ghana and Ivory Coast, thus, saving several million gallons of fresh water per annum. The gains of this technology are enormous considering the volume of money and other resources that are saved. For example, conventional toilets require sanitary vehicles for the evacuation of septic tanks when they are filled up. This becomes unnecessary since the wastes are now digested on-site. The cost of these trucks, which amounts to several hundreds of thousands of US dollars, is thus saved. Also, the need to constantly pipe water to the utilities is reduced due to the *in situ* recycle of water. Land conservation is another gain of this initiative considering that the digesters which are constructed using reinforced concrete are buried below the ground. This replaces the several square kilometers of land needed for sewage treatment at conventional waste treatment plants. Again the cost of energy is reduced when using fossil fuel (diesel) from between US$ 0.60–0.70 cents/kilowatt hour to between US$ 0.04–0.14 cents/kilowatt hour when using biogas [52,67].

In the water and food production sector, it is estimated that agriculture accounts for the use of 900 km³ of water, which is 80% of all abstracted groundwater in the world [68]. Much of this water is lost to evaporation and ground seepage because most of the water is channeled through bare soil furrows, open canals and flooded plains. This wastage is being curtailed in countries such as Bangladesh, China, India, USA, Iran, and Pakistan where much irrigation activities are practiced through the use of precision water-saving technologies such as sprinkler and drip irrigation systems [68]. Thus, rather than investing in abstracting more groundwater—thereby jeopardizing ground water retaining structures/aquifers—more efficient use of already abstracted water is advocated. This approach, which ensures better crop yield, can be described as a soft-path approach to the management of groundwater resources. These examples demonstrate the wide applications of the trialog model and the soft path approach in resource management.

Local Case Study

Lagos State has been classified as the second fastest growing mega city in Africa and the seventh fastest growing city in the world [69,70]. With a current population of 21 million people and growth rate of 3.2%, the State has a population density of 20,000 person/km² in its built-up areas [2,71]. One of the challenges created in areas with high population density is transportation problems. In the first instance, constructing new roads or expanding existing roads can be very costly. In the second instance, the existing land constraint situation could adversely affect other aspects of the society if more road networks are constructed. Therefore, the appropriate management of existing infrastructure becomes an attractive alternative. This is known as the soft path approach to resource management. To solve the transportation problem, the Lagos Bus Rapid Transport (BRT) system was created in 2008. Although it is not the first BRT model in the world, the Lagos BRT is unique in that it was the cheapest in terms of implementation when all other factors such as population served and cost of implementation per kilometre are taken into consideration [72,73,74]. Details of the project included the purchase of 220 midi-buses, construction of bus parks/boarding stations, partitioning of existing roads using 400 mm high kerbs and/or paint markings for the BRT routes. At a total implementation cost of $37.4 million, this represents a fraction of what would have been needed to construct new road networks. Today, the BRT project transports 60 million passengers per annum and is self sustaining [72,74]. A survey revealed that 65% of those who benefitted from the Lagos BRT in 2009 had no personal vehicles, while 25% (1.5 million people) gave up the use of their cars to take advantage of the facility [74]. This translates into a lot of energy savings as well as prevention of vehicular emissions which could have been dispersed into the atmosphere. Further, the Lagos BRT succeeded because of the collaboration between all stakeholders (the government, transport consultants (researchers) and the society). The government provides the regulations through its agency, Lagos metropolitan area transport management agency, LAMATA. Government also procured 120 buses which it leased to private operators. The remaining 100 buses were procured directly by private operators. Other benefits provided by the BRT project included a 30% reduction in fares in comparison to existing transport schemes in the State. There was also a 40% and 35% reduction in travel time and queuing times respectively, besides safety and comfort [74]. A BRT television show with a weekly audience of five million people is broadcasted twice weekly to inform the public and to provide a platform for a much needed feedback system. This example further illustrates the dynamics of the trialog model.

Stakeholder Responsibilities

The presented case studies have highlighted the responsibilities of different parties in the trialog model. It has been shown that government can gain a great deal by partnering with research institutions/scientists and other members of society (the public, NGOs and Civil Societies) in solving society's problems. Each partner within the model, however, has roles and responsibilities. The government ought to provide regulatory and security services, while the scientific institutions/researchers have the responsibility of fashioning out solutions and serving in advisory capacity to both government and society. The private sector and NGOs (as part of society), in many cases, also provide infrastructural services for their employees and communities in the areas in which they operate. They could also serve as conduits through which government educates and communicates its policies on water, sanitation and energy to the larger society. On the other hand, the private sector can also be culpable in the degradation of the environment. Although it is nearly impossible not to generate waste in the process of production, the impact of such generated wastes on the environment can be minimized through appropriate law enforcement guided by advisory input from the scientific community. In addition, civil societies and organized unions could serve as checks on the activities of the private sector by making reports of observed pollution activities to the relevant government regulatory agencies.

CONCLUSIONS

This study has demonstrated that thousands of lives are being adversely affected as a direct result of environmental degradation arising from unsustainable development practices in some parts of Africa. While human development is highly dependent on the use of natural resources, human survival is much more dependent on the practice of sustainable and responsible resource utilization. A proper understanding of the concept of the trialog model as well as the "soft path" approach can help all stakeholders to effectively play their role in the drive towards sustainable human development. These two concepts advocate smooth working relationships among stakeholders as well as the appropriate use and management of resources rather than exploitation of more resources. The trialog model identifies government's primary role as regulatory. It also identifies the scientific community as the advisory party responsible for developing ways through which resources can be optimally exploited and appropriated for human use without repercussions on either humans or the environment. The third party in the trialog model (society) is the main beneficiary, as well as primary custodian of environmental capital. Although the proponents of the trialog model have adapted it to water resource management, it is the position of this study that the merits of the model could be extended to the governance of all other resources such as ambient air, forests, food and energy resources.

Furthermore, due to increasing developmental activities, as well as Africa's reputation, as the continent with the highest energy intensity, Africa has the potential of becoming a major contributor to global warming. Although Africa contributed just 3.6% to global carbon emission as at 2008 [48], the situation could worsen rapidly if adequate measures are not set in motion as early as possible. Therefore, research and development effort should be directed at creating alternative sources to replace the use fossil and solid fuels. Potential energy sources that could be explored for use in Africa include wind, geothermal, solar, hydropower and biogas. In the short-term, however, private sector operators of industries which currently generate more power than they can consume should be encouraged to sell such excess energy to their host communities. In addition, favorable policies should be offered as incentives to encourage private sector entrepreneurs who wish to invest in environmentally-friendly energy production. This could help reduce the number of people who depend on solid fuel as energy source.

ACKNOWLEDGMENTS

The authors would like to thank Tshwane University of Technology for funding this research. Oluseyi Ajayi is acknowledged for proof reading and language editing.

AUTHOR CONTRIBUTIONS

This research was designed by David O. Omole and the paper was jointly written by David O. Omole and Julius M. Ndambuki.

REFERENCES

1.   Dasgupta, P. The idea of sustainable development. *Sustain. Sci.* 2007, *2*, 5–11.

2.   Omole, D.O.; Isiorho, S.A. Waste management and water quality issues in coastal states of Nigeria: The Ogun State experience. *J. Sustain. Dev. Africa* 2011, *13*, 207–217.

3.   United Nations. Report of the World Commission on Environment and Development. Available online: http://www.un.org/documents/ga/res/42/ares42-187.htm (accessed on 30 July 2014).

4.   Montgomery, M.R. The Urban Transformation of the Developing World. *Science* 2008, *319*, 761–764.

5.   UN-Habitat. *State of the World'S Cities 2006/7*; United Nations: New York, NY, USA, 2006.

6.   Pacione, M. *Urban Geography: A Global Perspective*, 2nd ed.; Routledge, Taylor and Francis: Abingdon, UK, 2005.

7.   Daramola, A.; Ibem, E.O. Urban environmental problems in Nigeria: Implications for sustainable development. *J. Sustain. Dev. Africa* 2009, *12*, 24–45.

8.   UNPF. State of World Population 2009. Available online: http://www.unfpa.org/webdav/site/global/shared/documents/publications/2009/state_of_world_population_2009.pdf (accessed on 30 July 2014).

9.   Westoff, C.F. The implications of United Nations long-range population projections: Continuing rapid population growth. In *United Nations Department of Economic and Social Affairs. Population Division. World Population to 2300*; United Nations: New York, NY, USA, 2004.

10. Hazard, E.; de Vries, L.; Barry, M.A.; Anouan, A.A.; Pinaud, N. The Developmental Impact of the Asian Drivers in Senegal. *World Econ.* 2009, *32*, 1563–1585.

11. UN. Sustainability and equity: A better future for all. In *Human Development Report 2011*; United Nations Development Program: New York, NY, USA, 2011.

12. Green, R.K.; White, M.J. Measuring the benefits of home-owning: Effects on children. *J. Urban Econ.* 1997, *41*, 441–461.

13. Marsh, A.; Gordon, D.; Heslop, P.; Pantazis, C. Housing deprivation and health: A longitudinal analysis. *Hous. Stud.*2000, *15*, 411–428.

14. Wolf, C.G.; Schroeder, D.G.; Young, M.W. Effect of improved housing on illness in children under 5 years old in northern Malawi: Cross sectional study. *Br. Med. J.* 2001, *322*, 1209–1212.

15. United Nations. Official List of Millennium Development Goals Indicators. Available online: http://unstats.un.org/unsd/mdg/Host. aspx?Content=Indicators/OfficialList.htm (accessed on 30 July 2014).

16. Moe, C.L.; Rheingans, R.D. Global Challenges in Water, Sanitation and Health. *J. Water Health* 2006, *4*, 41–57.

17. Gasana, J. Water and Health. *Air Water Borne Dis.* 2014, *3*, 1–3.

18. UN-WOMEN. Facts and Figures. Available online: http://www.unwomen. org/en/news/in-focus/commission-on-the-status-of-women-2012/facts-and-figures (accessed on 30 July 2014).

19. WHO. *Domestic Water Quantity, Service, Level and Health*; WHO Document Production Services: Geneva, Switzerland, 2003.

20. RSC. Africa's Water Quality: A Chemical Science Perspective. A Report by the Pan Africa Chemistry Network. Available online: http://www.rsc. org/images/RSC_AWQ_PACN_Flyer_tcm18-176916.pdf (accessed on 30 July 2014).

21. UNICEF. Women and Hunger: 10 Facts. Available online: http://www. wfp.org/our-work/preventing-hunger/focus-women/women-hunger-facts (accessed on 30 July 2014).

22. Omole, D.O. *Water Quality Modelling: Case study of the Impact of Abattoir Effluent on River Illo, Ota, Nigeria*; LAP Lambert Academic Publishing GmbH & Co. KG: Saarbrücken, Germany, 2010.

23. UN. *Human Development Report 2013. The Rise of the South: Human Progress in a Diverse World*; United Nations Development Program: New York, NY, USA, 2013.

24. Arimah, B. Poverty Reduction and Human Development in Africa. *J. Hum. Dev.* 2004, *5*, 399–415.

25. AWV. The Africa Water Vision for 2025: Equitable and Sustainable Use of Water for Socioeconomic Development. Available online: http://

www.afdb.org/fileadmin/uploads/afdb/Documents/Generic-Documents/african%20water%20vision%202025%20to%20be%20sent%20to%20wwf5.pdf (accessed on 4 July 2014).

26. Harhay, M.O. Water Stress and Water Scarcity: A Global Problem. *Am. J. Public Health* 2011, *101*, 1348–1349.

27. Curry, E. Water Scarcity and the Recognition of the Human Right to Safe Freshwater. *Northwest. J. Int. Hum. Rights* 2010, *9*, 103–121.

28. Longe, E.O.; Omole, D.O.; Adewumi, I.K.; Ogbiye, A.S. Water Resources Use, Abuse and Regulations in Nigeria. *J. Sustain. Dev. Africa* 2010, *12*, 35–44.

29. UN WWAP. The World Water Development Report 3: Water in a Changing World. Available online: http://www.unesco.org/water/wwap/wwdr/wwdr3/ (accessed on 30 July 2014).

30. Gleick, P.H. Global freshwater resources: Soft path solutions for the 21st century. *Science* 2003, *302*, 1524–1528.

31. Gleick, P.H.; Allen, L.; Christian-Smith, J.; Cohen, M.J.; Cooley, H.; Herberger, M.; Morrison, J.; Palaniappan, M.; Schulte, P. *The World's Water Volume 7: The Biennial Report on Freshwater Resources*; Pacific Institute for studies in Development, Environment and Security: Washington, DC, USA, 2012.

32. Howsam, P. *Water Law, Water Rights and Water Supply (Africa)*; Department for International Development and Cranfield University: London, UK, 1999.

33. Olaosebikan, B. *Lagos State Water Corporation: A New Dawn in Water Supply*; Lagos State Water Corporation: Ijora, Nigeria, 1999.

34. Ajai, O. Law, Water and Sustainable Development: Framework of Nigerian Law', 8/1. *Law Environ. Dev. J.* 2012, *8*, 89–115.

35. Omole, D.O. Sustainable Groundwater Exploitation in Nigeria. *J. Water Resour. Ocean Sci.* 2013, *2*, 9–14.

36. UNEP. Report on Atmosphere and Air Pollution. African Regional Implementation Review for the 14th Session of the Commission on Sustainable Development (CSD-14). Available online: http://www.un.org/esa/sustdev/csd/csd14/ecaRIM_bp2.pdf (accessed on 30 July 2014).

37. Tanimowo, M.O. Air pollution and respiratory health in Africa: A review. *East Afr. Med. J.* 2000, *77*, 71–75.

38. Hopkins, J.R.; Evans, M.J.; Lee, J.D.; Lewis, A.C.; Marsham, J.H.; McQuaid, J.B.; Parker, D.J.; Stewart, D.J.; Reeves, C.E.; Purvis, R.M. Direct estimates of emissions from the megacity of Lagos. *Atmos. Chem. Phys.* 2009, *9*, 8471–8477.

39. Kampa, M.; Castanas, E. Human health effects of air pollution. *Environ. Pollut.* 2008, *151*, 362–367.

40. WHO. Fuel for Life: Household Energy and Health. Available online: http://www.who.int/indoorair/publications/fuelforlife.pdf (accessed on 3 September 2012).

41. Frolicher, T.L.; Winton, M.; Sarmiento, J.L. Continued global warming after CO_2 emissions stoppage. *Nat. Clim. Chang.* 2014, *4*, 40–44.

42. Satterthwaite, D. Cities' contribution to global warming: Notes on the allocation of greenhouse gas emission. *Environ. Urban.* 2008, *20*, 539–549.

43. Ogunsola, O.J.; Oluwole, A.F.; Asubiojo, O.I. Environmental impact of vehicular traffic in Nigeria: Health aspects. *Sci. Total Environ.* 1994, *147*, 111–116.

44. Margulis, S.; Paunio, M.; Acharya, A. Addressing Indoor Air Pollution in Africa: Key to Improving Household Health. 2006. Available online: http://www.unep.org/urban_environment/PDFs/IAPAfrica.pdf (accessed on 1 September 2012).

45. Barrios, S.; Bertinelli, L.; Strobl, E. Climatic change and rural–urban migration: The case of sub-Saharan Africa. *J. Urban Econ.* 2006, *60*, 357–371.

46. Seth, S. Inequality, Interactions, and Human Development. *J. Hum. Dev. Capab.* 2009, *10*, 375–396.

47. Nair, M. *Renewable Energy for Africa*; Institute for Environmental Security: Hague, The Netherlands, 2009.

48. USEIA. International Energy Outlook 2011. U.S. Energy Information Administration. 2011; Available online: www.eia.gov/ieo/pdf/0484(2011).pdf (accessed on 1 September 2012).

49. Ajayi, O.O. Assessment of utilization of wind energy resources in Nigeria. *Energy Policy* 2009, *37*, 750–753.

50. IEA. *Energy Statistics of non-OECD Countries 2000–2001*; International Energy Agency: Paris, France, 2003.

51. Karekezi, S.; Kithyoma, W. Rnewable Energy in Africa: Prospects and Limits. In Proceedings of the Workshop for African Energy Experts on

Operationalizing the NEPAD Energy Initiative, Dakar, Senegal, 2–4 June 2003.

52. Deichmann, U.; Meisner, C.; Murray, S.; Wheeler, D. The economics of renewable energy expansion in rural Sub-Saharan Africa. *Energy Policy* 2011, *39*, 215–227.

53. IEA. *Key World Energy Statistics*; International Energy Agency: Paris, France, 2011.

54. USAID. Property rights and governance: Nigeria. Available online: http://usaidlandtenure.net/sites/default/files/country-profiles/full-reports/USAID_Land_Tenure_Nigeria_Profile.pdf (accessed on 22 March 2014).

55. Taye, O.O.; Dada, M.O. Appraisal of Private Sector Involvement in Infrastructure Development in Lagos State, Nigeria. *Mediterr. J. Soc. Sci.* 2012, *3*, 399–412.

56. Anand, S.; Segal, P. What Do We Know about Global Income Inequality? *J. Econ. Lit.* 2008, *46*, 57–94.

57. Franks, T.; Bdliya, H.; Mbuya, L. Water governance and river basin management: Comparative experiences from Nigeria and Tanzania. *Int. J. River Basin Manag.* 2011, *9*, 93–101.

58. Knuppe, K.; Pahl-Wostl, C. A framework for the analysis of Governance Structures Applying to Groundwater Resources and the Requirements for the Sustainable Management of Associated Ecosystem Services. *Water Resour. Manag.* 2011, *25*, 3387–3411.

59. Hattingh, J.; Maree, G.A.; Ashton, P.J.; Leaner, J.J.; Turton, A.R. A trialogue model for ecosystem governance. *Water Policy* 2007, *9*, 11–18.

60. Turton, A.; Godfrey, L.; Julien, F.; Hattingh, H. Unpacking Groundwater Governance through the Lens of a Trialogue: A Southern African Case Study International Symposium on Groundwater Sustainability (ISGWAS). Available online: http://www.ehrn.co.za/publications/download/113.pdf (accessed on 29 July 2014).

61. Conley, H.A.; Dukkipati, U. Leading from Behind in Public-Private Partnerships ? In *An Assessment of European Engagement With the Private Sector in Development*; Centre for Strategic and International Studies: Washington, DC, USA, 2012; pp. 1–13.

62. Larson, H.J.; Heymann, D.L. Public Health Response to Influenza A(H1N1) as an Opportunity to Build Public Trust. *J. Am. Med. Assoc.* 2010, *303*, 271–272.

63. Yahya, M. Polio vaccines—"No thank you!" barriers to polio eradication in Northern Nigeria. *Afr. Aff.* 2007, *106*, 185–204.

64. Herizonte, B. Infrastructure in Latin America, Slow! Government Obstacle ahead. *Economist* 2006, *41*, 4–18.

65. Doyin, A. Solutions to International Challenges in PPP Model Selection: A Cross-Sectoral Analysis. In Proceedings of the Deloitte Research and the London School of Economics, London, UK, 13 March 2006.

66. Idan, J.A. Present Discards (Wastes) Management Situation In Ghana and its Solutions. In Proceedings of the African Roundtable for Sustainable Consumption and Production (ARSCP)/GEMS Roundtable on Wastes Management for Sustainable Development, Obafemi Awolowo University, Ile-Ife, Nigeria, 16–18 November 2011.

67. Ohno, Y. Sustainable Energy for the African Bottom Billion. Master's Thesis, The University of Tokyo, Tokyo, Japan, 2010.

68. Shah, T.; Burke, J.; Villholth, K. Groundwater: A global assessment of scale and significance. In *Water for Food, Water for Life: A Comprehensive Assessment of Water Management in Agriculture*; Earthscan: Colombo, Sri Lanka; International Water Management Institute: London, UK, 2007.

69. Aderogba, K.A. Global Warming and Challenges of Floods in Lagos Metropolis, Nigeria. *J. Acad. Res. Int.* 2012, *2*, 448–468.

70. Aluko, O.E. The impact of Urbanization on Housing Development: The Lagos Experience, Nigeria. *Ethipian J. Environ. Stud. Manag.* 2010, *3*, 64–74.

71. LASG. Population. Lagos State Bureau of Statistics. Available online: http://www.nigerianstat.gov.ng/pages/download/187 (accessed on 7 December 2013).

72. LAMATA. BRT/BFS. Available online: http://www.lamata-ng.com/brt.php (accessed on 7 December 2013).

73. Athanasios, M.; Papageorgiou, G.; Ioannou, P.; Aphamis, T. Planning for Effective Bus Rapid Transit Systems: A Scenario Simulation Modelling Based Approach. In Proceedings of the IFAC Symposium on Control in Transportation Systems, Sofia, Bulgaria, 12–14 September 2012; Volume 13, pp. 366–371.

74. Mobereola, D. *Africa's First Bus Rapid Transit Scheme: The Lagos BRT-LiteSystem*; The International Bank for Reconstruction and Development/ The World Bank: Washington, DC, USA, 2009.

Chapter 5

TACKLING AIR POLLUTION IN CHINA— WHAT DO WE LEARN FROM THE GREAT SMOG OF 1950S IN LONDON

Dongyong Zhang[1,2], Junjuan Liu[1], and Bingjun Li[1]

[1]College of Information and Management Science, Henan Agricultural University, 15 Longzi Lake Campus, Zhengzhou East New District, Zhengzhou, Henan 450046, China

[2]Center for International Earth Science Information Network, The Earth Institute, Columbia University, P.O. Box 1000 (61 Route 9W), Palisades, NY 10964, USA

ABSTRACT

Since the prolonged, severe smog that blanketed many Chinese cities in first months of 2013, living in smog has become "normal" to most people living in mainland China. This has not only caused serious harm to public health, but also resulted in massive economic losses in many other ways. Tackling the current air pollution has become crucial to China's long-term economic and social sustainable development. This paper aims to find the causes of the current severe air quality and explore the possible solutions by reviewing the current literature, and by comparing China's air pollution regulations to that of the post London Killer Smog of 1952, in the United Kingdom (UK). It is hoped that China will learn the lesson from the UK, and decouple its economic growth from the detrimental impact of environment. Policy suggestions are made.

INTRODUCTION

The extraordinary economic achievement of China in the past more than two decades has lifted around 500 million people out of poverty, however, this achievement has rested on heavily polluting industries, on burning coal for energy, and on an explosion in the number of cars. Research by the Chinese Academy of Social Sciences pointed out that the problem of haze and fog in China was hitting a record level, and China is currently suffering the worst air pollution problem since 1961 [1].

The problem of air pollution was first observed in the 1970s, with industrial emissions of sulphur dioxide (SO_2) and total suspended particulates (TSP). In the 1980s, acid rain was detected in major cities in the northern part of the country, and this was mainly caused by SO_2 from coal combustion, which accounts for more than 70% of the fuel consumption in China. In the 1990s, the number of vehicles on roads increased very rapidly, especially in medium-sized and large cities. In Beijing alone, the number of vehicles increased by a factor of 10, from 0.5 million in 1990 to 5 million in 2012. In addition, the emission factor (the amount of pollution emitted by one car) in China is much higher than in developed countries because China has much lower emission standards for automobiles [2]. Thus, the drastic rise in the number of vehicles and rapid development of industries in cities has led to worsening air quality, and concentrations of nitrogen oxides (NO_x) and particulates are especially high. High levels of ozone concentration were frequently observed in summer and fall in several big cities. Between 1981 and 2001, ambient concentrations of TSPs in Beijing were more than double of China's National Annual Mean Ambient Air Quality Standard of 200 $\mu g/m^3$ [3] and five times the level that prevailed in the United States before the passage of the Clean Air Act in 1970 [2]. Since 2000, more than 92% of residents in China have been exposed to PM2.5 (the tiny particles which penetrate deep into the lungs and give rise to asthma, cancer, heart trouble, *etc.*) concentration exceeding 10 $\mu g/m^3$, the exposure rate increased to 98% in 2012, while during the same period the proportion of population that were exposed to PM2.5 concentration exceeding 10 $\mu g/m3$ in the UK decreased from 76% to 21%, and from 60% to 16% for the United States of America [4] (see Figure 1). The same study indicated that China is ranked as one of the bottom three countries, based on air quality, just above Nepal and Bangladesh (ibid.).

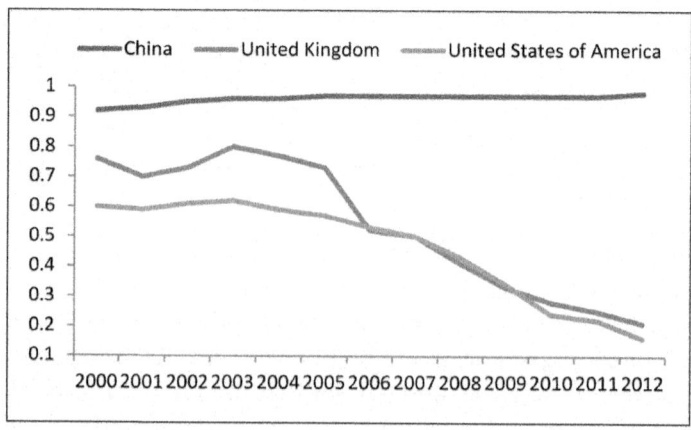

Figure 1. Proportion of the population exposed to a PM2.5 concentration of 10 $\mu g/m^3$.

Along with smog becoming a fixture of life, the public concern exploded in 2013 as the skies over many Chinese cities grayed. In Beijing, levels of PM2.5 were stuck at hazardous levels for weeks in early 2014 and peaked at 35 times the World Health Organization's (WHO) recommended limit. A total of 51.8% of days in 2013 were unhealthy or worse [5] (see Figure 2). The capital's 21 million residents put on face masks, kids were kept indoors, and social networks exploded with complaints about the heavy blanket of smog. In Harbin, a city in northern China, a dense wave of smog began on 20 October 2013, the day when the coal-powered district heating system started, visibility was reduced to below 50 meters in parts of Harbin and below 500 meters in most of the neighboring Jilin province. It was reported that the daily particulate levels in parts of Harbin municipality were more than 40 times the WHO recommended maximum level. All highways in the surrounding Heilongjiang province were closed. All primary and middle schools, and the airport, were closed for three days in Harbin. The hospital reported a 23% increase in admissions for respiratory problems. While the situation in Beijing and Harbin is alarming, they are not unique; many cities in Northern China, which is home to several of the world's most polluted cities [6], air quality is especially poor. Even state-supported media provided surprisingly critical coverage of the crisis and many foreign media called it an "airpocalypse".

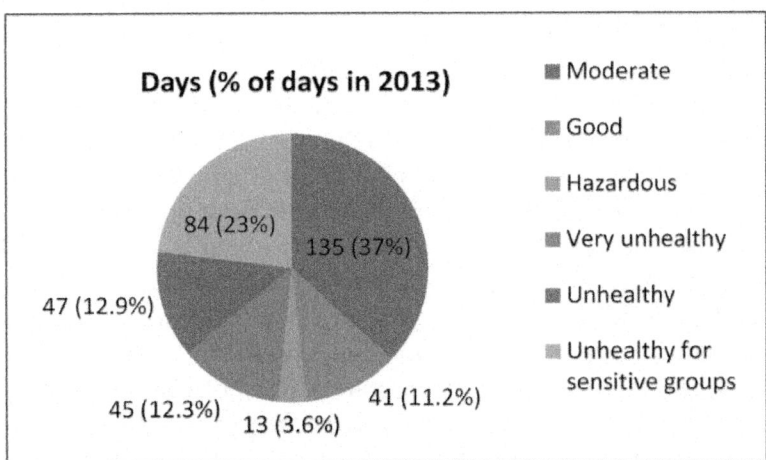

Figure 2. Beijing pollution levels: days in 2013 at Air Quality Index ratings.

The air pollution problems in China now are actually in parallels with the killer smog in London 60 years ago in December 1952 when a heavy motionless layer of smoky, dusty fumes from the region's millions of coal stoves and local factories appeared in the sky. This smog brought traffic and people to a standstill. Hospital admissions, pneumonia reports, applications for emergency bed service, and mortality followed the peak of air pollution. The mortality stayed high in the following two months, and it was suggested that 12,000 unexplained deaths during this period were owing to the smog. This smog became known as the "Great Smog" because of its lethality and the unprecedented public reactions to it [7].

Not accidentally, in 1952, the UK's stage of economic development, its real GDP per capita, was approximately the same as China's is today [8]. Along with the change of policy, London's air quality improved soon after 1952. China's Premier Li Keqiang declared the country's "war on smog" at the opening of China's National People's Congress, on 5 March 2014. China's winning or losing of this war will not only affect its own long-term sustainability but will make the most important turning point in the global scene on climate change. Looking at the lessons of London's air pollution catastrophe and how subsequently the situation was improved may cast some useful lessons for China. This paper aims at looking at the causes of smog in China and find possible solutions by review the current literature and by comparing the air pollution policies in China and post 1952 in the UK.

SOCIAL AND ECONOMIC CONSEQUENCES OF THE SMOG IN CHINA

The stark consequences for each Chinese resident brought by the deadly pollution are obvious, and various studies have been trying to put a measurement on the economic and social impacts that the smog has had in China. One of the focuses of the studies has been the human health effects of the smog. It was reported by the World Bank [6] that the economic burden of premature mortality and morbidity associated with air pollution was estimated to be 157.3 billion Yuan in 2003, or 1.16% of its GDP. A more recently research found that air pollution has caused the loss of more than 2.5 billion years of life expectancy in China, and, because of air pollution, linked diseases, such as cardiovascular disease and lung cancer due to high consumption of coal in northern China, life expectancy there was 5.5 years shorter than that in southern China [9]. Another recent study found that if the world took action to reduce greenhouse gas emissions, more than 500,000 lives could be saved globally each year, and the air and health quality benefits for East Asia alone would add up to between 10 and 70 times the cost of reducing emissions by 2030 [10]. The same study also found that the health benefits of taking action to curb climate change were especially striking for China, with its large population now exposed to some of the worst pollution in the world (ibid.). A study focusing on the babies born in the southwestern Tongliang county, China just before a local coal-fired power plant was shut down a decade ago found that air pollution led to genetic changes that these babies had significantly lower level of a protein that is crucial to brain development in their cord blood than those conceived later, and poorer learning and memory skills were also found in these kids when tested at the age of two [11].

Non-human-health-related impacts have also been evaluated by some studies. Matus *et al.* [12] applied the method developed for the US and Europe to China to estimate the socio-economic costs generated by air pollution in China. It was found that air pollution in China has created a substantial burden to its economy and the estimated ozone and PM concentrations beyond background levels have led to US$ 16 billion to US$ 69 billion (or 7% to 23%) loss of consumption and US$ 22 billion to US$ 112 billion (or 5% to 14%) loss of welfare in China's economy. The World Bank [6] reported that, although the impacts of pollution on natural resources (agriculture, fish, and forests) and manmade structures (e.g., buildings) were estimated to be lower in economic terms, acid rain, caused mainly by increased SO_2 emissions due to increased fossil fuel use, causes over 30 billion Yuan in damages to crops, primarily vegetable crops (about 80% of the losses). This amounts to 1.8% of the value of agricultural output. Damage to building materials in the South imposed a cost of seven billion Yuan on the Chinese economy in 2003. In addition, although the impact of smog on forests in China has not been quantified due to lack of monitoring data in remote areas, clear widespread damage has been observed to trees, forests around sites of particularly high pollution [13].

A new research conducted by China Agricultural University indicated that if the smog persists, China's agriculture will suffer conditions similar to a nuclear winter because the air pollutants, by adhering to greenhouse surfaces, cut the amount of light inside by about 50% and severely impeding photosynthesis, the process that helps plants convert light into life-sustaining chemical energy [14].

The smog has also changed the way people live their lives. At a cost of five million Yuan, the International School of Beijing has built two domes that enclose the entire school outdoor areas so as to protect the students from the bad air. Students can play and exercise, year round, in the domes without being polluted by the smog [15]. According to a report produced by Hurun Research Institute, China is losing its most important residents to smog. The report shows that in 2013, 64% of China's rich (those with wealth above $1.6 million) were either immigrating to another country or planning to, a rise of 60% from 2011 [16]. The same report indicated that the pollution and food safety were the second biggest reason for the rich to leave China, after the general desire for security and financial well-being.

CAUSES OF SMOG

Coal Burning

It was generally recognized that one of the main causes of the smog in December 1952 was the widespread use of coal [7]. In China, coal is also regarded as the number one source of air pollution [17]. The grey sky in Beijing is mainly due to coal burning, vehicle exhaust, climate and geographical environment, and other factors, such as crop stubble burning and firework [1]. From 1950 to 1980, the Chinese government provided free coal for home and office winter heating systems for anyone living north of the Huai River and Qin Mountain range (see Figure 3). The reason to choose Huai River and Qin Mountain as the divide was that Chinese government could not afford supplying free heating to all of China and the Huai River follows the January 0 °C (32 °F) average temperature line. Northern cities received free unlimited central heating between November and March. In contrast, central heating facilities in the Southern areas did not exist until recently when some private heating providers came to the market. Indeed, it is widely recognized that it feels colder in parts of southern regions that are closer to the Huai River (*i.e.*, Nanjing, Shanghai, and Chengdu) in winter as the temperature is only slightly higher than the north but not as high as in the far south. After the marketization, China's free central-heating system has been replaced by heavily subsidized central-heating system, but the supply is still only for northern China.

China's heating system is coal-based and technically inefficient [18]. Heat has been provided by coal-fired heat-only boilers or combined heat and power generators, which are inefficient in energy usage compared to electricity, gas, and oil heating systems in the industrial countries [19]. There are normally one or two heating providers in a city, hot water travels certain distance from the heating provider to each household, which causes substantial energy loss. The incomplete combustion of coal in these boilers leads to the release of at least three kinds of pollutant—total suspended particulates (TSP), CO_2 and NO_x. The amount of pollution produced varies depending on the type of coal used, which is relevant to the geographical area that the coal is produced. It is estimated that in China, coal combustion (including industrial and domestic use) is responsible for 87% of SO_2 and 76% of NO_x emissions [18].

Figure 3. Northern and southern China division according to domestic heating policy.

In additional to domestic use of coal, China's industrial use of coal is high too. China's ever-growing army of coal-fired power plants, of which there are currently more than 2300, including iron and steel and cement factories. According to the newly released international data from US Energy Information Administration), China's energy mix in 2012 contained 68% coal. China's total coal use grew by 325 million tons in 2011, accounting for 87% of the 374 million ton global increase in coal use. China's coal consumption in 2012 is 4.7 times of the US's, and almost 60 times that of the UK's. China accounts for 47% of global coal consumption in 2012—almost as much as the entire rest of the world combined for the same year (see Figure 4). Although China has made great progress in the investment in renewable energy, and its renewable electricity growth was double that of the US from 2010 to 2012, China's reliance upon coal is predicted to keep growing [20].

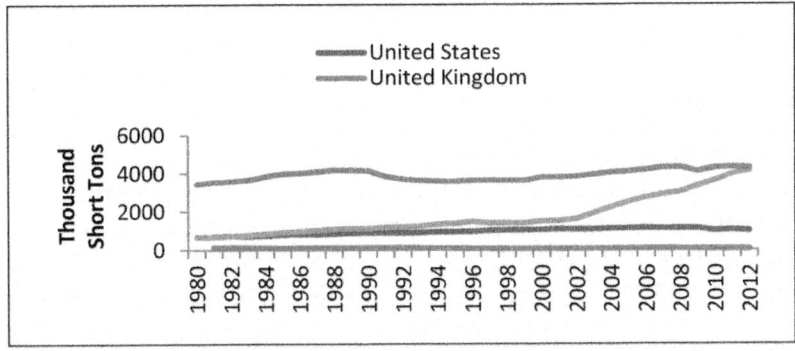

Figure 4. China's coal consumption comparing to the rest of the world.

Industrial Pollution

Research by the Chinese Academy of Science found that industrial pollution is the biggest source of PM2.5 problems, which cause Beijing's smog. Secondary inorganic aerosols, sulphates and nitrates are responsible for 26% of Beijing's PM2.5, followed by industrial production and coal burning at 25% and 18%. Soil dust accounted for 15%. The rest of the pollution comes from heavily industrialized neighbouring provinces and burning of trash, and Beijing's 5.5 million cars were found to be responsible for just 4% of the smog [21]. Zhao et al. [22] sampled PM2.5 of different seasons in Beijing, Tianjin, and Hebei, China, and analysed its chemical composition and seasonal variation from 2009 to 2010. The results indicated that the PM2.5 pollution was severe in Beijing, Tianjin, and Hebei, and the meteorological condition and pollution sources that affect aerosol pollution were season dependent in all areas. Thus, the concentrations of PM2.5 and its major chemical species were also season dependent i.e., lower in spring and summer and higher in autumn and winter in all sampling areas.

Vehicle Emissions

Vehicle emissions have also become a key source of air pollution in China. Research has indicated that the causes of air pollution in China's megacities have shifted from coal-burning only to a mix of coal burning and vehicle emissions [23]. Although Zhang et al. [21] determined that motor vehicle emissions account for less than 4% of Beijing's hazardous PM2.5 readings, Wang [24] claimed that pollutants from vehicles contribute to more than 22.5% of PM2.5 in Beijing, and the management of Beijing Environmental Protection Bureau claimed the collective rate of vehicle emissions contribution to PM2.5 should be between 20% and 30% because the secondary inorganic aerosol (which contributes to more than one quarter of the PM2.5) is largely from car emissions.

Other Factors

Smog in China can also be caused by other activities. Crop stubble burning is a method used frequently by farmers to deal with agricultural waste. About 40% of large portions of crop residues are used as biomass fuels in Hebei Province and Inner Mongolia, 55% in Heilongjiang Province and Liaoning Province, 70% in Tianjin City and Beijing City [1]. Firework is also a cause of the increase of PM2.5. According to Li [25], fireworks increased the level of PM2.5 by 80 times during the traditional Chinese New Year in 2012. A total of 260,000 cases of fireworks were used during the Chinese New Year in 2013 [1].

TACKLING THE SMOG

Policy Changes in London after the Smog

Although it was slow, the government responded to the Great Smog, Clean Air Act 1956, which was later amended and extended to Clean Air Act 1968, was passed by the Parliament of the United Kingdom. Social, economic, and technological changes were introduced to help reduce smoke and SO_2 emission:

- "Smokeless zones" and "smoke control areas" were set up in some towns and cities in which only smokeless fuels could be burned.
- Sources of household heating were shifted from coal to cleaner coals, electricity, and gas.
- Measures were also introduced to relocate power stations away from cities, and the height of some chimneys were required to be increased.

The act had dramatic effect on reducing the use of coal for domestic use. In 1952, coal supplied 61% of London's energy needs, and 28% was for household use alone. The household use of coal started to decline in 1957 and was totally replaced by environmentally friendly energy supplies such as electricity, oil, gas, *etc.*, and SO_2 generated from household heating were significantly reduced [7].

Standards for ambient air quality were set for the first time in the Mayor's Air Quality Strategy and come into effect in 1980. The Environment Act 1995 required government to produce a national air quality strategy containing standards and objectives, and measures to achieve the objectives. It also established a system of local air quality management and required local authorities periodically to review and assess the current and future quality of air in their areas.

At the end of 1980s, along with the change of industrial structure and source of energy, road traffic became the major cause of NO_2 in London's air, accounting for approximately 60% of emissions in London [7]. A few measures were introduced to road traffic: investment in the public transport network, congestion charging, appropriate planning, and other mechanisms. To reduce emissions of vehicles already on the road, the following measures were taken:

- In the short term, targeting emissions reductions from the most polluting vehicles, mainly heavier diesel vehicles, such as buses, coaches, goods vehicles, waste vehicles, and taxis.

- Increasing the take-up of newer, cleaner vehicles and technologies.
- Increasing the take-up of cleaner fuels.
- Introducing low emission zones in London, which would exclude the most polluting vehicles from specified areas.
- For the long term, promoting zero emission forms of transport, such as hydrogen fuel cell vehicles.

Measures were also taken to reduce the emissions from air travel, from industry and construction. These measures effectively reduced air pollution in London. From 1950 to 2000, the annual average smoke and SO_2 concentrations in London were reduced by 95% and 98% respectively [7].

China's Way to Clean Air

Law Enforcement

The prolonged, severe smog in first months of 2013 has shone a spotlight on the need for strong environmental regulations in china and prompted the government to move forward with a number of new environmental policies and laws at national level, some of which have been languishing in the proposal stage for years:

- China's Air Pollution Prevention and Control Law (first adopted in 1995, first amended in 2000, and reopened in 2013 for new update)
- National 10 Measures (adopted in June 2013)
- Action Plan for Air Pollution Prevention and Control (Action Plan, adopted in September 2013)
- Environmental Protection Law (first adopted in 1989, reopened for update in 2012, and passed in April 2014)
- Performance Assessment Measures for Air Pollution Prevention and Control Action Plan (for trial implementation) (adopted in April 2014)

    Regional environmental policies:

- Building a Beautiful Tianjin Programme (adopted in August 2013)
- Action Plan for Air Pollution Prevention and Control in Jing-Jin-Ji and Surrounding Areas—Details for implementation (adopted in September 2013)
- Action Plan of Beijing City for Clean Air 2013–2017 (adopted in September 2013)
- Implementation Details of Action Plan for Air Pollution Prevention and Control in Hebei Province (adopted in September 2013)

- Action Plan of Shanghai City for Clean Air 2013–2017 (adopted in October 2013)
- Action Plan of Guangdong Province for Air Pollution Prevention and Control 2014–2017 (adopted in February 2014)
- The 2014 Action Plan for Air Pollution Prevention and Control in Shan'xi Province (adopted in April 2014)
- Action Plan of Henan Province for the Blue Sky (adopted in May 2014)

National 10 Measures were disclosed in a strongly-worded statement to prevent and control air pollution. With 35 sub-sections, the 10 measure areas including the following:

- Strengthening the comprehensive efforts to reduce emission of multi-pollutants.
- Promoting industrial upgrading and restructuring.
- Accelerating companies' technology upgrading and increasing their technological innovation capacities.
- Accelerating energy restructuring to increase clean energy supply.
- Enforcing energy-saving and environmental protection as market entrance requirements.
- Imposing strict approval requirements for new investment projects regarding energy-saving and environmental protection, and restricting the investment of industries with high energy consumption and high pollution levels in environmentally vulnerable areas.
- Improving legal framework, implementation and enforcement.
- Establishing monitoring, alerting and emergency response systems for air pollution episodes.
- Defining responsibilities for environmental protection between the government, private sector and the public.
- Establishing a regional coordination mechanism to coordinate regional environmental governance.

To effectively implement the National 10 Measures, the Action Plan for Air Pollution Prevention and Control (Action Plan) was released by China's State Council in September 2013. The Action Plan is based on the 12th Five-Year Plan on Air Pollution Prevention and Control in Key Regions which was released in 2012 but contains much more stringent requirement. The objective of the Action Plan is to improve air quality and reduce the air pollution episodes in China, especially the three key regions Beijing-Tianjin-Hebei area (Jing-Jin-Ji), Yangtze River Delta (YRD), and Pearl River Delta (PRD). It focuses not only on pollution targets, but also on industrial restructuring, industrial location, and technological innovation, as well as stronger governance. It is enforced by linking industrial project approvals to EIA with energy audits, and linking the pollution reduction targets with senior officials' performance evaluations [23]. It was regarded as a milestone document in China's air pollution control history, and in the history of environmental management and governance [26].

Different targets are set for Jing-Jin-Ji, YRD and PRD areas for improvement by 2017 and where the Jing-Jin-Ji region is the most stringently targeted:

- For all second and third tier cities, annual average concentration of PM10 should be reduced by at least 10% comparing to the 2012 level, and the number of days with clean air should be increased.

- For the three key regions, the annual average concentration of PM2.5 should be reduced by 25%, 20% and 15% respectively.

- For Beijing, the annual average concentration of PM2.5 should be controlled at 60 $\mu g/m^3$ level.

Specific targets for coal consumption and vehicle emissions control were set in the Action Plan:

- By 2017, the proportion of coal in total energy consumption in China will be reduced to 65% of that in 2012, while the proportion of non-fossil energy consumption will be increased to 13%. The three key regions shall make efforts to achieve negative growth of total coal consumption, and replace coal with natural gas for coal-fired boilers, industrial furnaces, and self-sustained coal-fired power stations.

- By 2015, all yellow-labelled vehicles (vehicle registered before the end of 2005) shall be phased out in the three key regions. By 2017, all yellow-labelled vehicles will be phased out nationwide.

- Cleaner gasoline and diesel will be provided step by step, and by 2017 China V gasoline and diesel will be used for all vehicles in China nationwide.

The Action Plan put forward new incentives for the local government to improve their local air quality. Firstly, central government will have to disclose 10 best and 10 worst air quality cities monthly. Secondly, meeting targets for PM2.5 in the three key regions and PM10 in other key areas other than the three key regions is considered as compulsory for provinces and is part of the performance evaluation indicators for provincial leaders. In addition, regional collaboration mechanisms in Jing-Jin-Ji and YRD are established with the participation of provincial governments and relevant central ministries in the region. Environmental protection and meteorological agencies will set up air pollution monitoring and alert system.

In response to the Action Plan, local governments have released their local action plans. Beijing and Hebei set up detailed targets including the tonnes of coal to be reduced. Beijing laid down more than 40 measures on vehicle emissions control, in particular, setting vehicle cap at the 6 million levels by 2017.

China's Air pollution Prevention and Control Law, which has not been amended since 2000 was reopened by China's Ministry of Environmental Protection (MEP) earlier in 2013, but the amending process has been delayed, indicating the huge conflict of interests among affected groups. However China's first Environmental Protection Law, which dates back to 1989, and after two years of debate, had its first amendment in April 2014, approved by Standing Committee of National People's Congress The new law "sets environmental protection as the country's basic policy" [27]. A prominent change in this revision is that the administration of the environment has been given a legal framework, which means that in China now there is a stronger and more official system of duty. The new law will remove limits on fines for polluters, which are currently so low that many factories prefer to pay them than take long-term anti-pollution measures. It encourages "studies on the impact environmental quality causes on public health, urging prevention and control of pollution-related diseases" [27] and promises greater powers for environmental authorities and stricter punishments for polluters, for example it allows authorities to detain company managers for 15 days if they do not complete environmental impact assessments or ignore warnings to stop polluting. The amendment also includes a chapter on information disclosure stating that citizens have the right to obtain information about the environment.

The Gap between Legislation and Implementation

Although the new Environmental Protection Law is praised to have "provides smooth and orderly channels for the public to make appeals on environmental subjects" and have the potential to become the cornerstone for China's "war on pollution" [27], some environmental groups say that China's greatest environmental problems arise from a gap between legislation and implementation [28]. Can the new law fill in this gap? "Implementation presents problems in all the new provisions, there will be technical obstacles, problems in matching the new rules with existing situation on the ground, and issues when the changes in the law alter the power of interest groups", stated Xu Nan, Deputy Editor of China Dialogue a website that monitors environmental issues in China. The hope is that these revisions will give the

Ministry of Environmental Protection the power it needs to enforce these new, high-minded eco-protection laws.

In addition, critics say the targets set for Beijing's annual average concentration in the Action Plan are conservative, because 60 μg/m³ as annual average concentration of PM2.5 in 2017 is still higher than the 35 μg/m³ limit set up by the National Ambient Air Quality Standard (issued in 2012 and to be implemented in 2016) for annual average PM2.5 concentration. Furthermore, targets for the annual average PM2.5 concentration in cities in the three key regions, except Beijing, are unclear, as they require percentage reductions in PM2.5 based on a base year. However, the annual average concentration of PM2.5 in those cities in the base year (2012) has not been reported and may not even exist because PM2.5 was only widely monitored and assessed in China since 2013.

Democracy and Pollution

Environmental degradation is the result of the single-minded pursuit of economic strength without democratic accountability [29]. The development of Taiwan's environmentalism is closely synchronized with successive stages in transition to democracy, and environmental protests in Taiwan in the 1980s over pollution caused by the island's industrialization improved people's sense of political efficacy and pushed Chiang Ching-Kuo to lift martial law in 1987, which ushered in Taiwan's process of democratization [30]. Although it is bold to say mainland China will follow Taiwan's step and the social unrest caused by mainland China's pollution could bring about democracy, it is evident that high profile protests against various polluting industries and the recent shocking air quality in mainland Chinese cities have helped place the environment at the center stage in Chinese politics. Chinese official media used to describe the country's pollution problem as a necessary but temporary consequence of its economic transformation, but the heavy smog in early 2013 has made poisonous air the become lead item in the prime-time news, broadcast continuously by the state broadcaster China Central Television. The report was not just extensive, but also critical.

Why the change of tone? "Leaders are aware that people can wait 20 years or more for democracy but they cannot wait that long for clean air," says the editor of China Dialogue, which covers environmental issues in China [31]. Companies in China often ignore the environmental laws because of loose enforcement, weak penalties and a prevailing attitude of "I have money hence I can do anything". However, several successful public protests in recent years against polluting projects give China the hope of achieving democracy through environmental issues. The new Environment Protection Law proposed the mechanism of transparency promotion, which includes requiring companies to monitor and report real-time pollution data, clearly specifying criminal penalties for those who evade such monitoring systems or forge monitoring data [32]. In addition, the new law forbids improperly operating pollution prevention equipment and holds government agencies responsible for disseminating information publicly [33]. The new Law also moves close to democracy by permitting civil society organizations to initiate public interest lawsuits on behalf of citizens.

Public Awareness

Chapter 20 of China's Agenda 21 endorses the goal of increasing public participation in environmental governance [34]. Without the involvement of the public, it would be difficult to restructure institutions to serve necessary ecological rationalities such as the circular economy and sustainable development. Traditional top-down pollution control approaches are increasingly becoming inadequate in enforcing regulations. Based on the number of environmental tips received on official environmental protection hotline 12369, a recent report shows public tip-offs regarding environmental issues are increasing in China, indicating the rising public awareness of the problem. With 26% increase of the number of total environmental tips received on 12369 from 2012 to 2013, more than 70% of the tips concerned airborne pollution in 2013. According to the ministry's environmental emergency and accident investigation center, airborne pollution has become the hottest environmental topic among the public [35]. The same report found that public awareness of the pollution is higher in areas with lower level of air quality. Hebei Province, the most polluted province in China, set up a system in 2013 to reward residents who report environmental illegal behaviours, which obviously have contributed to increasing public awareness of environmental pollution in the province and nationwide.

However, to what extent would public awareness contribute to environmental protection? Van Rooij [36] found that Chinese pollution victims continue to face formidable institutional barriers to effective action including cost, slowness, lack of impartiality, corruption, legal and scientific knowledge, lack of intermediary institutions, *etc.*, while the polluting companies have enough resources and power to shed their responsibility and get away from punishment due to their close connection to the local government. The China General Social Survey of 2006 asked the respondents who considered themselves as pollution victims whether they had taken action on the most severe occasion. Only a quarter had done so. Of the types of actions taken, contacting the polluter or the government departments were the two dominant strategies. Of the 169 people who took action, 63% were unhappy with the result, and the type of action taken seemed to make little difference. Only 4% said they went to the news media and only 1% went to court. Of the ones who did not take actions, 43% thought it would be of no use, 11% said they it would be too much trouble and pricy to take action, 19% did not what to do, and 22% said they could put up with the problem so the did not even thought of taking action [37]. Hence, it remains sceptical how the emergence of public participation in a country without tradition of participatory democracy. Zhao [34] found that the major constraints of public participation of environmental impact assessment (EIA) are the limited extent of public participation in EIA, limited access to information, limited import of the public in decision-making, and limited access to judicial redress and remedy.

DISCUSSION AND RECOMMENDATIONS

Air pollution has been killing people since the dawn of industrialization, and China's air pollution problem is no worse than London's great smog in 1952 and Japan's smog of the 1960s. However, more is known about its risk now and global warming raises the stakes: China overtook the US as the biggest source of greenhouse gases in 2006 and has put the globe on a path to exceed UN targets for the rise in the Earth's temperature. However, with coal still providing about 65% of China's energy, it will take years to reverse its dependence on polluting fossil fuels. As the issue threatens to expand from national environmental activism to a public obsession and global attention, China's leaders have pledged to be transparent about the air pollution in China and vowed not to repeat mistakes that cost them public trust during the SARS outbreak in 2003 and the tainted milk scandal in 2008, and a "war on smog" was declared by Chinese Premier Li Keqiang in March, 2014. Just like what happened in London after the killer smog in 1952, China updated Environmental Protection Law and issued relevant action plan and measures as weapons to help China win the "war".

However, it takes more than just weapons to win a war. The past track record of China in implementing environmental laws have not given people too much confidence in seeing China winning the war. With similar causes of air pollution such as coal burning, vehicle emissions, China can certainly learn from London's experience in the 1950s in tackling the smog problem. However, there are also differences in the scale of the events: London's smog primarily affected the city, while the recent smog events covered large areas over China. In addition, anomalies in the meteorological conditions in China, as well as wind blow dust deserts contributed to the severity of the recent smog events [38]. Therefore, the lessons learned from London may not be well suited for China. The aggressive pollution control measures during 2008 Beijing Olympic Games and Paralympic Games (20 July–17 September), such as factory closures and traffic control had effectively improved the air quality in Beijing. However, these pollution control actions were temporary, and the pollution came back more aggressively. Thus, a more consistent follow-up to the policies maybe more important than stringent policies themselves.

As long as China still worships GDP, China will have to burn more coal to power most of its factories. In fact, a few big emitters account for most of the pollution. For example, the eight dirties enterprises in Shandong and Hebei Province (the dirtiest province in China) produce 37 and 30 times emissions of NOx respectively. In Hebei, 21 power plants discharge more than 90% of the emissions. If these plants actually adhered to the new limits set in the updated law, their emissions would fall by more than half. However, with the "Chinese dream" to realized, the GDP worship will continue. The giant state-owned polluters will still be pressured by the local government to boost GDP, PM2.5 levels will continue to surge. The main obstacle is that the biggest polluters tend to be the best connected politically. They ignore the rules because it is cheaper to do so and because they can. In contrast, the implementation agencies are weak and overstretched, and lack the ability to enforce all their well-meaning rules and standards. Hence, although it is crucial to be equipped with stringent environmental laws and regulations, shorten the gap between legislation and implementation, increase transparency and democracy regarding environmental pollution, and raise public awareness, Chinese government will have to focus on the core of the problem—tackling the pollution, and stop its GDP worship.

ACKNOWLEDGMENTS

The authors would like to thank the two reviewers for their useful comments on the paper, and thank Alex de Sherbinin and Marc Levy at CIESIN, Columbia University for their valuable suggestions. The authors would also like to thank the financial support from Henan Science and Technological Innovation Talent Fund (094100510013), Henan Soft Science Fund (092400440039), and from Humanities and Social Science Research Project (2012-DZ-046) and Soft Science Project (13A630500) of the Educational Department of Henan Province.

AUTHOR CONTRIBUTIONS

Dongyong Zhang contributed to the completion of the main body of the manuscript, Bingjun Li and Junjuan Liu contributed to all the literature review in Chinese language and part of the literature review in English.

REFERENCES

1. Wang, K.; Liu, Y. Can Beijing fight with haze? Lessons can be learned from London and Los Angeles. *Nat. Hazards*2014, *72*, 1265–1274.

2. Policy and Global Affairs (PGA); Chinese Academy of Engineering (CAE); Chinese Academy of Science (CAS); National Academy of Engineering (NAE); National Research Council (NRC). *Urbanization, Energy, and Air Pollution in China: The Challenges Ahead—Proceedings of a Symposium*; National Academies Press: Beijing, China, 2005.

3. Bi, X.; Feng, Y.; Wu, J.; Wang, Y.; Zhu, T. Source apportionment of PM10 in six cities of Northern China. *Atmos. Environ.* 2007, *41*, 903–912.

4. Hsu, A.; Emerson, M.; Levy, M.; de Sherbinin, A.; Johnson, L.; Malik, O.; Schwartz, J.; Jaiteh, M. The 2014 Environmental Performance Index. Yale Center for Environmental Law and Policy: New Haven, CT, USA. Available online: http://issuu.com/yaleepi/docs/2014_epi_report (accessed on 31 May 2014).

5. Beijing Municipal Environmental Monitoring Center (BMEMC). The annual average concentration of PM2.5 of 2013 was 89.5 microgram/ m^3. Available online: http://www.bjmemc.com.cn/g327/s922/t1913.aspx (accessed on 31 May 2014).

6. World Bank. *Cost of Pollution in China: Economic Estimates of Physical Damages*; The World Bank: Washington, DC, USA, 2007.

7. Greater London Authority. *50 Years on—The struggle for Air Quality in London since the Great Smog of December 1952*; Greater London Authority: London, UK, 2002.

8. Measuringworth. Logarithm of UK real GDP per capita (2008 pounds), China real GDP per capita (thousands of 2005 yuans). Available online: http://www.measuringworth.com/graphs/graph_1.php (accessed on 31 May 2014).

9. Chen, Y.; Ebenstein, A.; Greenstone, M.; Li, H. Evidence on the impact of sustained exposure to air pollution on life expectancy from China's Huai River policy. Available online: http://pluto.huji.ac.il/~ebenstein/ PNAS-2013-Chen-1300018110.pdf (accessed on 31 May 2014).

10. West, J.; Smith, S.J.; Silva, R.A.; Naik, V.; Zhang, Y.; Adelman, Z.; Fry, M.M.; Anenberg, S.; Horowitz, L.W. Co-benefits of mitigating global greenhouse gas emissions for future air quality and human health. *Nat. Clim. Change* 2013, *3*, 885–889.

11. Tang, D.; Lee, J.; Muirhead, L.; Li, T.Y.; Qu, L.; Yu, J.; Perera, F. Molucular and neurodevelopmental benefits to children of closure of a

coal burning power plant in China. *PLoS One* 2014, *9*, e91966.

12. Matus, K.; Nam, K.; Selin, N.E.; Lamsal, L.N.; Reilly, J.M.; Paltsev, S. Health damages from air pollution in China.*Global Environ. Change* 2012, *22*, 55–66.

13. United Nations Environment Programme (UNEP). Forests suffer from air pollution. Available online: http://www.unep.org/vitalforest/Report/ VFG-19-Forests-suffer-from-air-pollution.pdf (accessed on 7 June 2014).

14. Kaiman, J. China's toxic air pollution resembles nuclear winter, say scientists. *The Guardian*, 2014. Available online: http://www. theguardian.com/world/2014/feb/25/china-toxic-air-pollution-nuclear-winter-scientists (accessed on 31 May 2014).

15. McKirdy, E. China looks for blue-sky solutions as smog worsens. *CNN World*, Available online: http://www.cnn.com/2014/02/24/world/asia/ beijing-smog-solutions/ (accessed on 31 May 2014).

16. Hurun report. The Chinese Millionaire Wealth Report 2013. Available online: http://up.hurun.net/Humaz/201312/20131218145315550.pdf (accessed on 31 May 2014).

17. Facts and Details. Air pollution in China. Available online: http:// factsanddetails.com/china/cat10/sub66/item392.html(accessed on 31 May 2014).

18. Almond, D.; Chen, Y.; Greenstone, M.; Li, H. Winter heating or clean air? Unintended impacts of China's Huai River policy. *Am. Econ. Rev.* 2009, *99*, 184–190.

19. Jiang, Y. Promoting Chinese energy efficiency. *China Dialogue*, 2007. Available online: https://www.chinadialogue.net/article/show/single/ en/1119 (accessed on 31 May 2014).

20. Larson, E. China's growing coal use is world's growing problem. *Climate Central*, 2014. Available online: http://www.climatecentral.org/blogs/ chinas-growing-coal-use-is-worlds-growing-problem-16999 (accessed on 31 May 2014).

21. Zhang, R.; Jing, J.; Tao, J.; Hsu, S.C.; Wang, G.; Cao, J.; Lee, C.S.L.; Zhu, L.; Chen, Z.; Zhao, Y.; *et al.* Chemical characterization and source apportionment of PM2.5 in Beijing: Seasonal perspective. *Atmosp. Chem. Phys.* 2013, *13*, 7053–7074.

22. Zhao, P.S.; Dong, F.; He, D.; Zhao, X.J.; Zhang, X.L.; Zhang, W.Z.; Yao, Q.; Liu, H.Y. Characteristics of concentrations and chemical compositions for PM2.5 in the region of Beijing, Tianjin and Hebei, China. *Atmosp. Chem. Phys.* 2013, *13*, 4631–4644.

23. Institute for Global Environmental Strategies (IGES). *IGES Policy Report: Major Development in China's National Air Pollution Policies in the Early 12th Five-Year Plan Period*; IGES Policy Report No. 2013-02 2014. IGES: Kanagawa, Japan ISBN: 978-4-88788-163-1. Available online: http://pub.iges.or.jp/modules/envirolib/upload/4954/attach/Major_Developments_in_China's_Air_Pollution_Policies_March2014.pdf (accessed on 12 August 2014).

24. Wang, S. Motor vehicle contributes to more than 22.2% PM2.5 rate. *Jinghua Times*, 2014. Available online: http://epaper.jinghua.cn/html/2014-01/03/content_52877.htm (accessed on 15 June 2014).

25. Li, Q.D. New Year Eve's firecrackers raise PM2.5 rate for 80 times. *Jinghua Times*, 2012. Available online: http://epaper.jinghua.cn/html/2012–01/24/content_755044.htm (accessed on 15 July 2014).

26. Clean Air Alliance of China (CAAC). State Council Air Pollution Prevention and Control Action Plan. In *China Clean Air Updates*; 2013; The original document is in Chinese, issued by State Council on 10 September, 2013 (Document No. GUOFA[2013]37).

27. Xinhua News Agency. Xinhua Insight: China Declares War Against Pollution. Available online: http://news.xinhuanet.com/english/special/2014-03/05/c_133163557.htm (Accessed on 9 August 2014).

28. Roney, T. Will China's new environmental protection law make a difference? *The Diplomat*, 25 April 2014. Available online: http://thediplomat.com/2014/04/will-chinas-new-environmental-protection-law-make-a-difference/ (accessed on 31 May 2014).

29. Arrigo, L.G. The environmental nightmare of the economic miracle: Land abuse and land struggles in Taiwan. *Bull. Concerned Asian Sch.* 1994, *26*, 21–44.

30. Ho, M. Environmental movement in democratizing Taiwan (1980–2004): A political opportunity structure perspective. In *East Asian Movements*; Broadbent, J., Brokeman, V., Eds.; Springer: New York, NY, USA, 2011; pp. 283–314.

31. Burkitt, L.; Spegele, B. Beijing fog prompts state media to shift tone. *Wall Str. J.* 2013. Available online: http://online.wsj.com/news/articles/SB10001424127887324595704578241640520226304 (accessed on 31 May 2014).

32. Environmental Protection Law of the People's Republic of China. Article 63. 2014.

33. Environmental Protection Law of the People's Republic of China. Article 65. 2014.

34. Zhao, Y. Public participation in China's EIA regime: Rhetoric or reality? *J. Environ. Law* 2010, *22*, 89–123.

35. Wu, W. Public awareness rises over air pollution. *China Daily*, 7 June 2014. Available online: http://www.chinadailyasia.com/news/2014-06/07/content_15139103.html (accessed on 31 May 2014).

36. Van Rooij, B. The people *vs.* pollution: Understanding citizen action against pollution in China. *J. Contemp. China* 2010,*19*, 55–77.

37. Munro, N. Profiling the victims: Public awareness of pollution-related harm in China. *J. Contemp. China* 2014, *23*, 314–329.

38. Huang, K.; Zhuang, G.; Wang, Q.; Fu, J.S.; Lin, Y.; Liu, T.; Han, L.; Deng, C. Extreme haze pollution in Beijing during January 2013: Chemical characteristics formation mechanism and role of fog processing. *Almos. Chem. Phys. Discuss.*2014, *14*, 7517–7556.

Chapter 6

THEORETICAL INVESTIGATIONS ON MAPPING MEAN DISTRIBUTIONS OF PARTICULATE MATTER, INERT, REACTIVE, AND SECONDARY POLLUTANTS FROM WILDFIRES BY UNMANNED AIR VEHICLES (UAVS)

Nicole Mölders[1,2], Mary K. Butwin[1,2], James M. Madden[1,2], Huy N. Q. Tran[1,3], Kenneth Sassen[1,4], Gerhard Kramm[1,5]

[1]Geophysical Institute, University of Alaska Fairbanks, Fairbanks, USA

[2]Department of Atmospheric Sciences, College of Natural Science and Mathematics, University of Alaska Fairbanks, Fairbanks, USA

[3]Bingham Entrepreneurship & Energy Research Center, Utah State University, Vernal, USA

[4]Department of Atmospheric Science, University of Utah, Salt Lake City, USA

[5]Engineering Meteorology Consulting, Fairbanks, USA

ABSTRACT

Evaluated Weather Research and Forecasting model inline with chemistry (WRF/Chem) simulations of the 2009 Crazy Mountain Complex wildfire in Interior Alaska served as a testbed for typical Alaska wildfire-smoke conditions. A virtual unmanned air vehicle (UAV) sampled temperatures, dewpoint temperatures, primary inert and reactive gases and particular matter of different sizes as well as secondary pollutants from the WRF/Chem results using different sampling patterns, altitudes and speeds to investigate the impact of the sampling design on obtained mean distributions. In this experimental design, the WRF/Chem data served as the "grand truth" to assess the mean distributions from sampling. During frontal passage, the obtained mean distributions were sensitive to the flight patterns, speeds and heights. For inert constituents mean distributions from sampling agreed with the "grand truth" within a factor of two at 1000 m. Mean distributions of gases involved in photochemistry differed among flight patterns except for ozone. The diurnal cycle of these gases' concentrations led to overestimation (underestimation) of

20 h means in areas of high (low) concentrations as compared to the "grand truth." The mean ozone distribution was sensitive to the speed of the virtual UAV. Particulate matter showed the strongest sensitivity to the flight patterns, especially during precipitation.

INTRODUCTION

In recent years, unmanned air vehicles (UAVs) have attained increasing attention from environmental scientists as UAVs permit measurements in hazardous air space (e.g. over wildfires) and/or over difficult to access remote areas [1] (e.g. Interior Alaska). The reasons are manifold. UAVs are much cheaper to purchase and deploy than manned aircrafts. In addition, the logistics for a UAV flight campaign have much shorter timeframes than planning an aircraft field campaign. This fact is especially critical in the research on phenomena that occur irregularly (e.g. volcanic eruptions) and for applications.

Over the last decades, communities in the boreal taiga have grown, and they are expected to grow further in the future [2]. In the sparsely populated, but wildfire-prone boreal taiga, a dense air-quality monitoring network would be required to provide public health advisory on smoke rolling into town from upwind wildfires. A recent study [3] showed that a network would require randomly distributed sites to provide the most representative data distribution. The installation and maintenance of such networks over the length of a fire season are expensive, especially in difficult to reach places. Furthermore, data are only required in case of a wildfire in a community's upwind region. This means data demand exists on an irregular basis over a limited area far outside of town and for a limited time. Performing measurements by flying UAVs in hard to reach or inaccessible regions may be cheaper than installing and maintaining a monitoring network in complex, permafrost-underlain, undeveloped terrain.

Due to its continental location, Interior Alaska summers are dry and warm with calm winds. Most of its summer precipitation is from convection and thunderstorms [4]. Thus, this region is prone to wildfires [4] - [7]. Due to the low population density [2], the road network is sparse. Any existing (meteorological) sites are biased to the road system consisting of one south-north highway and three west-east highways over an area as large as 20% of the contiguous US. Due to this remoteness, barely any monitoring of air quality exists outside the city limits of the Fairbanks metropolitan area, which is the only conurbation in Interior Alaska. Thus, this conurbation could benefit from UAV measurements for air-quality advisory when wildfires exist in its upwind [8].

In atmospheric sciences, UAVs have been deployed for measuring meteorological fields. A recent study [9], for instance, evaluated the atmospheric boundary layer (ABL) parameterizations of the advanced research Weather Research and Forecasting (WRF) model [10] for calm and gravity-wave conditions by UAV-collected temperature, relative humidity and wind profiles up to 3 km altitude above ground. The UAV has shown its value for examining the temporal evolution of the ABL including mesoscale features like subsidence inversions. The authors concluded that UAVs have the potential to close the observational gap for investigating relevant physical processes like mountain-induced gravity waves in the ABL. During the Verification of the Origins of Rotation in Tornadoes Experiment 2 (VORTEX2), for instance, UAVs took in-situ measurements in the rear flank downdraft and gust front to examine the thermodynamics in the lifecycle of super cells [11].

In geology, some studies focused on volcanic heat and gas emissions [12] [13]. One study, for instance, used a UAV at La Fossa Vulcano, Italy during April 2007 for the remote sensing of sulfurdioxide (SO_2) fluxes by ultraviolet and infrared spectrometers, and for measuring the carbondioxide (CO_2) to sulfurdioxide ratio. This study focused on the temporal rather than spatial aspect of the release.

In the above studies, the UAVs collected data over a small area and for phenomena that lasted comparatively short in time (e.g. formation of inversion, downdraft, gust front) and/or with the purpose of research. However, when the intent is to use data for air-quality advisory and to fly UAVs instead of installing and maintaining a monitoring network, it has to be examined whether UAVs can provide reliable mean spatial distributions of air-quality relevant quantities. Air-quality advisory namely relates to the National Ambient Air Quality Standard (NAAQS) that is defined for time ranges of 1 h to 24 h depending on the pollutant [14]. Furthermore, application of UAVs in air-quality advisory requires several hours for data collection to cover the area in the downwind of a wildfire in the upwind of a settlement. Furthermore, some of the chemical species undergo reactions thereby altering the chemical composition of the propagating smoke plume.

Our feasibility study examined whether UAVs could provide the spatial distributions of air-quality relevant information desired for air-quality advisories. To achieve our goal, we turned to numerical modeling and applied the analysis method by [3] described in Section 2. In our study, evaluated model data from a wildfire simulation served as a dataset representative of wildfire-smoke conditions ("grand truth") from which a virtual UAV sampled data. Section 3 presents an evaluation of the model data. In Section 4, the mean distributions derived from the sampled data were compared to the mean distribution according to the model data, which served as "grand truth." Here, the sensitivity of mean distributions derived from sampled data to three different flight patterns, at three different heights and at three different speeds is discussed. The conclusions end with recommendations for setting up UAV flight plans.

EXPERIMENTAL DESIGN

Reference Data "Grand Truth"

Model Setup

To investigate the impact of UAV flight patterns, speeds and altitudes on temporal mean spatial distributions derived from the sampled data, we ran WRF/Chem [15] [16] with the advanced research WRF dynamics solver [10] to obtain a four-dimensional physico-chemically consistent high-resolution dataset. We used the following physical and chemical packages. The Rapid Radiative Transfer Model [17] served to determine long-wave radiation under consideration of multiple absorption bands of atmospheric trace gases and cloud optical depth. The Goddard shortwave scheme [18] calculated diffuse and direct solar radiation under consideration of the ozone and cloud properties calculated inline [16].

A modified version of the Grell-Dévényi cumulus ensemble scheme [19] served to parameterize the impacts of the multiple mass fluxes from sub-grid scale convective clouds on the vertical profiles of temperature, moisture and wind. Cloud microphysical processes on the resolvable scale considered co-existance of super-cooled water (rainwater, cloud water) and ice, graupel and snow [20]. Interaction of clouds, radiation, and chemistry were considered as in [21] [22].

The processes in the ABL were parameterized using the Eta model Mellor-Yamada-Janjić schemes [23] [24]. These schemes, among other things, calculated the turbulent kinetic energy, buoyancy and shear in the ABL and free atmosphere.

The Rapid Update Cycle land-surface model [25] determined the exchange of momentum, heat, and matter at the Earth-atmosphere interface. It predicted the soil temperature, and soil water/ice conditions at six depths under consideration of frozen ground physics. Furthermore, it considered a one-layer canopy, fractional snow cover, snow depth, snow temperature and snow density, as well as surface albedo.

The gas-phase chemistry mechanism [26] considered 14 stable species, four reactive intermediates and three abundant stable species for inorganic chemistry, 26 stable species and 16 peroxy radicals for organic chemistry. The photolysis rates for 21 photochemical reactions were calculated inline as a function of wavelength, temperature, species, and absorption cross-section [27]. Dry deposition of trace gases followed [28] with the modifications for Alaska by [29].

The Modal Aerosol Dynamics for Europe [30] [31] calculated aerosol physics. Herein, the particle-size distribution from the submicron to coarse mode was parameterized by two log-normal modes in accord with [32]. The Secondary ORGanic Aerosol Model described the secondary aerosols formation by low volatility processes and gas-to-particle conversion [33]. Aerosol-removal processes by sedimentation and washout as well as some aqueous phase reactions were considered [16].

Emission Data

Emissions from biomass burning were created with the so-called PREP-CHEM-SRC emission processor [22] [34]. Information on the locations and daily advancement of fires stemmed from the Moderate Resolution Imaging Spectroradiometer (MODIS) wildfire database [35]. Emission rates and species depended on fuel maps [36] and area burned [16].

Anthropogenic emissions were generated from the Emission Database for Global Atmospheric Research (EDGAR) emission inventory, which provides annual emissions of greenhouse and precursor gases on a $1° \times 1°$ grid [37]. Emissions of SO_2, nitrogenoxides ($NO_x = NO + NO_2$, i.e. nitric oxide and nitrogen dioxide), carbonmonoxide (CO), particulate matter of 2.5 mm or less in diameter ($PM_{2.5}$) and 10 mm or less in diameter (PM_{10}) as well as volatile organic carbons (VOC) were allocated depending on the weekday, and hour of the day using Alaska-specific allocation functions [38]. The split for $PM_{2.5}$ and VOCs followed [29].

Biogenic emissions were calculated inline depending on land-use/cover following [39] [40]. Furthermore, this scheme included NO emissions by soil bacteria as a function of soil conditions.

Simulations

The model domain of interest covered the atmosphere over Interior Alaska centered at 65.57°N, 145.9°W with 110 ´ 100 grid-points of 4 km increment to 100 hPa (Figure 1). The initial and boundary conditions were downscaled from the $1° ´ 1°$ and 6 h resolution global final analysis data [41] of the National Centers for Environmental Prediction.

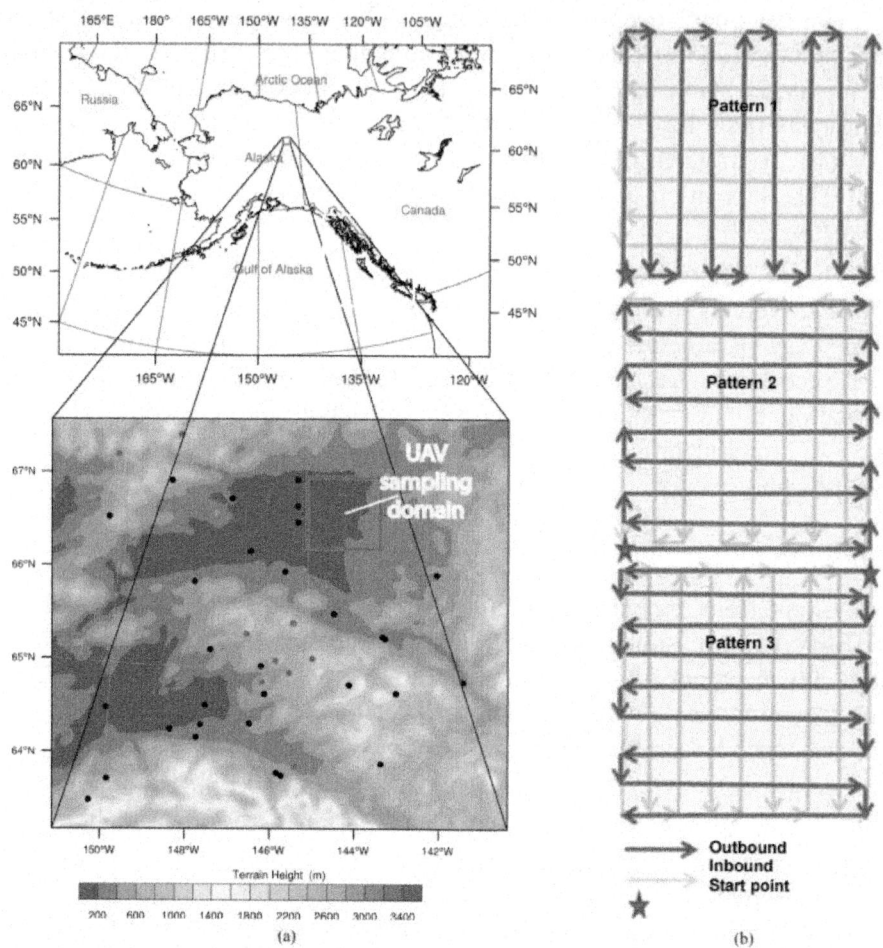

Figure 1. (a) Location of the WRF/Chem domain on which the evaluation was performed with location of the area sampled by the virtual UAV (magenta box). The black, blue, and red dots indicate the locations of the surface meteorological sites from the RAWS, ASOS, and SNOWTEL network, respectively; (b) Flight patterns for sampling by the virtual UAV.

The simulation started on August 3, 2009 0000 UTC with Alaska background concentrations. The chemical concentrations from the first two days of the simulation were excluded from the UAV sampling to allow for spin-up of the chemical fields. WRF/Chem was run in forecast mode for August 3 to 10, 2009. The meteorology was re-initialized every five days, while the chemical fields of the previous day served as initial data for the next.

Evaluation

To assess whether the WRF/Chem data represented a realistic dataset, we used data from 33 surface meteorological sites. The performance in predicting 2 m air temperatures, 2 m dewpoint temperatures, 10 m wind speeds and directions was quantified in terms of bias (simulated vs. observed), root-mean-square error (RMSE), standard deviation of error (SDE), and correlation-skill score (R).

Data of $PM_{2.5}$ from three sites in the Fairbanks metropolitan area served to assess WRF/Chem's performance in capturing the temporal evolution in this area and in case of 1-in-3-days data, the order of magnitude. We omitted calculation of spatio-temporal means for the following reasons: (a) All data were from the same area in the domain. (b) In this area, notable anthropogenic emissions occurred which was not the case anywhere else. (c) Too few data existed for a meaning full statistic. (d) Our study focused on wildfire smoke.

To assess the performance in predicting the height and vertical extension of the smoke plume, cross-sections of WRF/Chem predicted PM_{10} were compared qualitatively to Cloud-Aerosol Lidar and Infrared Pathfinder Satellite Observations (CALIPSO) level 1B backscatter and depolarization data. We used the backscatter data to assess orientations, sizes, and shapes of aerosols through the linear depolarization ratio (LDR) following [42] - [45]. The LDR is a ratio of backscattering powers in the perpendicular to parallel polarization planes [42].

In the interpretation of the CALIPSO data, we used the same considerations and thresholds as [46] [47]. In theory, perfectly spherical particles show no backscattering in the perpendicular, or orthogonal plane, while irregular shapes cause perpendicular backscattering [43] [48]. Generally, smoke particles are spherical [49] producing near zero perpendicular backscatter and marginal depolarization [50] [51]. Fresh smoke and smoke layers at high altitudes in the troposphere show depolarizations of 3% and 5%, respectively [48]. Aged smoke yields slightly elevated depolarization values because of coagulated particles and/or included soil particles [49] [52]. The upper parts of smoke layers have depolarization values of around 6%, while comparatively lower depolarization exists below this layer [50] [53]. Irregular-shaped particles like smoke particles cause some orthogonal backscatter [54].

UAV Virtual Sampling

The Crazy Mountain Complex fire of 2009 served as a testbed (Figure 1). While August 5 showed a fully developed smoke plume, August 6 presented a case wherein the passage of a cold front removed the smoke. August 7 to 10 permitted us to examine the rebuilding of the wildfire-smoke plume [8].

Following [3], the evaluated WRF/Chem data served to represent the atmospheric conditions during a wildfire event. These data are referred to as "grand truth" hereafter. The virtual UAV assumed in this study was a Scan Eagle. The Scan Eagle has a cruising speed of 111 $km \cdot h^{-1}$ and a theoretical flight time of 24 h without payload. Since any load including fuel reduces flight duration, we assumed 20 h flight duration as the best-case scenario. The virtual UAV sampled air temperature, dewpoint temperature, CO, SO_2, NO, O_3, $PM_{2.5}$ and PM_{10} from the model data using the different flight designs described below.

The area that the virtual Scan Eagle can cover within 20 h of sampling encompasses about 60 km × 60 km [8]. We applied the flight patterns as in [8]. The default flight pattern (FP1) was flying at cruise-speed at 200 m altitude. Herein the UAV started at 0000 UTC in the southwest corner of the sampling domain flying 60 km north then turning east for 4 km, turning south for 60 km, and again turning east for 4 km (Figure 1). The virtual UAV repeated this pattern until it reached the northeast corner of the sampling domain. Then the virtual UAV flew 60 km west, 4 km south, 60 km east, 4 km south and so on until reaching its original start point. Flight pattern 2 (FP2) assumed the same starting point, but with the long west-east/east-west scans first followed by the long south-north/north-south scans. The third flight pattern (FP3) started in the northeastern corner of the sampling domain flying the long south-north/north-south legs first followed by the long west-east/east-west legs. Three different velocities were examined: stall speed (72 $km \cdot h^{-1}$), cruise-speed (111 $km \cdot h^{-1}$), and maximum speed (148 $km \cdot h^{-1}$). Sampling occurred at 200 m, 500 m and 1000 m height. These heights correspond to the lowest level safe for UAVs traveling over the complex terrain of Interior Alaska, a level located in about the middle of the ABL and in the upper/around the top of the ABL, respectively. Since for air-quality advisory purposes the 200 m level is of greatest interest, the discussion of results mainly focused on this height.

The authors are well aware that sampling frequency, i.e. the number of readings per time unit, differ among instruments for the various quantities mentioned above. For simplicity of our theoretical investigations, we assumed a frequency of one reading per second for all instruments mounted on the virtual UAV.

Since the WRF/Chem output data were recorded at one-hour intervals on a lattice with 4 km increment, the sampled quantities were interpolated in time and space between available WRF/Chem data. The field quantities were collected under consideration of the UAV's speed and wind speed as a distance-time weighted mean between the values at the grid-cell in which the UAV was located and the nearest grid-cells along the flight path, and between the values of the past and next WRF/Chem recording at these grid-cells. A grid-cell was sampled for the duration that the UAV flew in the grid-cell. Consequently, more data were collected within a grid-cell where the UAV faced headwind than in a grid-cell with tailwind. For instance, at zero wind speed and cruise speed, for a grid-cell that is not located at the boundaries of the 60 km ´ 60 km sampling domain, about 130 data were collected on the outbound and inbound paths each, i.e. in total 260 data. As it is obvious from Figure 1, for the grid-cells at the boundaries of the sampling domain, the UAV collected data on the short legs as well. This means these grid-cells were sampled four times (2 times 4 km, and 2 times 2 km). The grid-cells at the corners of the sampling domain were sampled four times for 2 km each. In the above example of zero wind speed, the virtual UAV would collect about 390 and 260 data, respectively.

Based on all the data sampled during the flight, a 20 h mean was calculated for the 60 km ´ 60 km sampling area. Furthermore, based on all data sampled within 20 h in a 4 km ´ 4 km grid-cell, a 20 h mean was calculated for each grid-cell to obtain the 20 h mean distribution within the sampling area. Such 20 h distributions were calculated for each day from August 5 to 10 for each sampled quantity.

We compared these distributions to the distribution of 20 h means calculated from the WRF/Chem data ("grand truth"). Differences in chemical field quantities were expressed in terms of normalized mean bias (NMB), and fractional mean bias (FB) in accord with [55] [56]. The latter weights positive and negative biases equally, while the former avoids inflation due to the range of observed concentrations [56].

In case of flying at maximum and stall speed, the UAV finished sampling the area in less and more than 20 h, respectively. In the plots, we showed the mean distributions obtained for sampling at different speeds no matter of how long it took the virtual UAV to cover the entire sampling area. Thus, be aware that plots show 20 h mean distributions for all sampling designs except the minimum and maximum speed scenarios. In the calculation of the statistics, we used sampled and "grand truth" 20 h mean distributions no matter whether the entire domain was already sampled (in case of stall speed) or sampling took less than 20 h (in case of maximum speed).

RESULTS AND DISCUSSION

Evaluation

Capturing the meteorology is central to air-quality forecasts [57]. In the evaluation, we considered all data available in the WRF/Chem domain to demonstrate that WRF/Chem provided a realistic four-dimensional consistent dataset of an Interior Alaska wildfire.

The obtained skill scores have similar magnitude as those of other WRF studies in high latitudes [7] [58] [59]. In our study, WRF/Chem captured the temporal evolution of 2 m temperature acceptably (Figure 2(a)). The overall bias, RMSE, SDE, and correlation-skill score were 2.7°C, 4.9°C, 4.1°C and 0.74, respectively. The diurnal course was dampened by about 3°C. Both minimum and maximum temperatures were higher than observed, i.e. the model atmosphere was overall too warm. The spatial variability of simulated 2 m temperatures was less than observed due to the smoothing of the landscape by use of a grid-cell mean terrain height and dominant land-cover and soil type. These assumptions have impact on the flux densities of heat, and water vapor and hence temperature and the formation of convective clouds [60] [61].

WRF/Chem captured the temporal behavior of 2 m dewpoint temperatures acceptably (Figure 2(b)) with overall bias, RMSE, SDE and correlation skill score of ~0°C, 3.6°C, 3.6°C, and 0.70, respectively. These performance skills fall in the range of other subartic WRF studies [7] [59] [62] [63].

On average over all sites, WRF/Chem captured the temporal evolution of 10 m wind speed acceptably (Figure 2(c)). At the beginning of the simulation, gusts related to thunderstorms, channeling effects and fire-related winds led to huge variability in observed wind speeds.

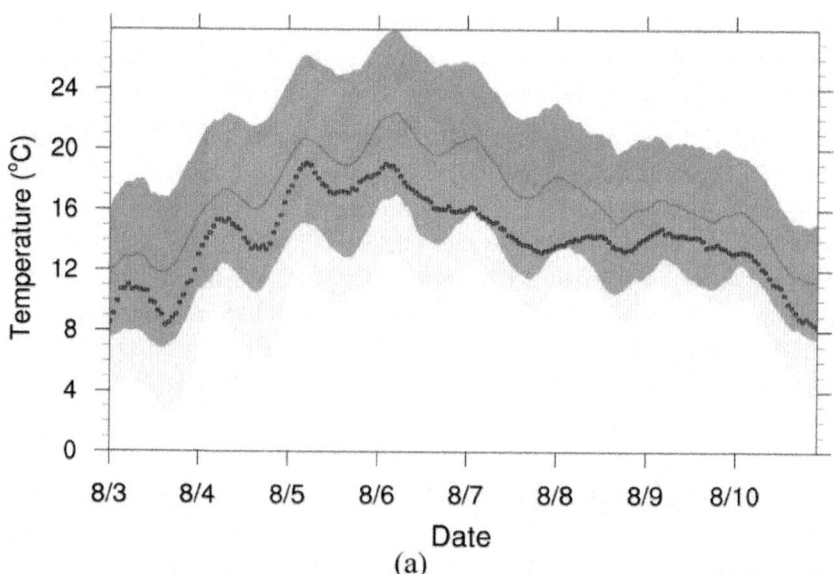

(a)

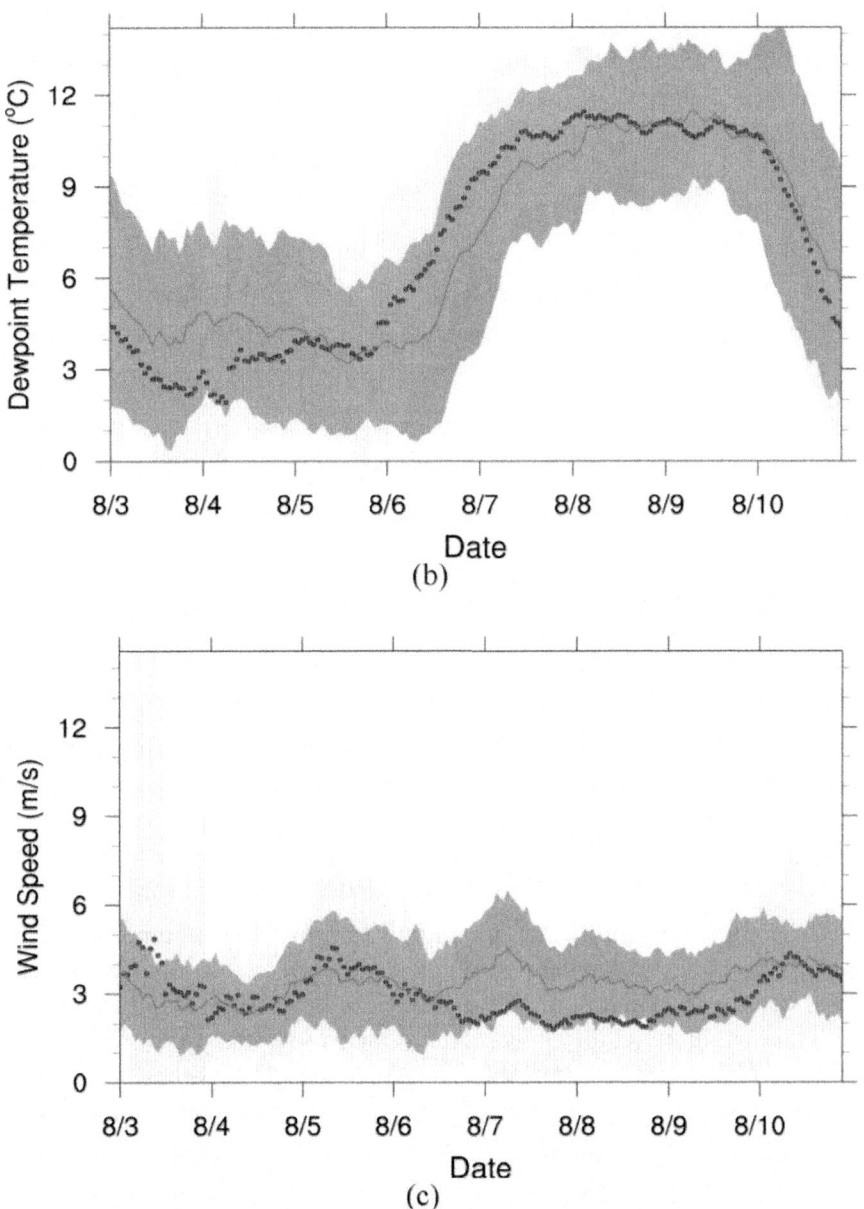

(b)

(c)

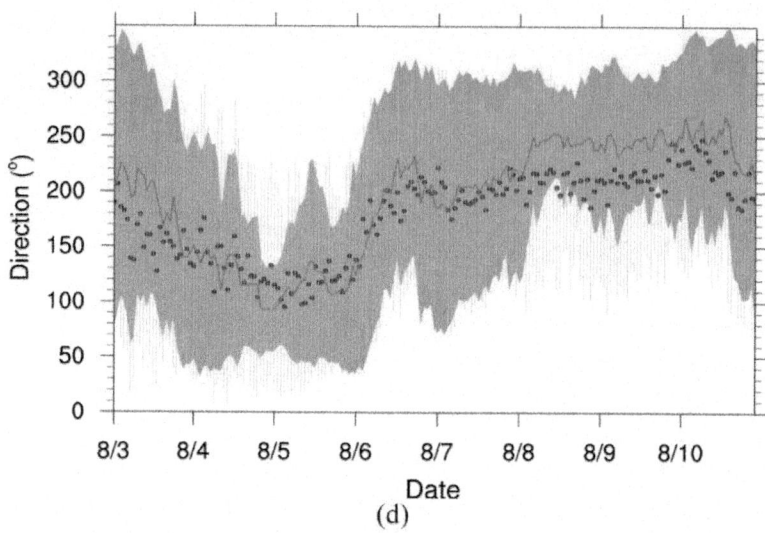

Figure 2. Spatial means and spatial standard deviations of hourly (a) 2 m air temperatures, (b) 2 m dewpoint temperatures, (c) 10 m wind speeds, and (d) 10 m wind directions as obtained from WRF/Chem (red line and orange shading, respectively) and the observations (blue dots and gray bars, respectively) using all sites with available data.

During frontal passage, simulated and observed wind speeds were slightly offset, but agreed better in magnitude than during calm conditions. WRF/Chem overestimated 10 m wind speeds during the stagnant conditions in the middle of the episode. Overall, these shortcomings yielded a positive bias of 0.1 m·s⁻¹ with RMSE and SDE of 6.2 m·s⁻¹ and low correlation.

The complex terrain caused the high variability of wind directions (Figure 2(d)) as the valleys channel the wind. Furthermore, mesoscale circulations also affected wind direction. In the model, valleys were of subgrid scale for which the spatial variability in wind direction was about a factor of two smaller than according to the observations. WRF/Chem best captured wind direction between August 5 and 8 and showed too strong westerly component otherwise.

Simulated PM$_{2.5}$ at the State Office building in Fairbanks showed similar general temporal behavior as the observations (Figure 3). However, simulated concentrations were much smaller in magnitude than observed. Due to errors in temporal offsets canceling each other out, mean fractional bias and normalized mean error amounted to 10% and 20%, respectively. The average fractional difference was 42%. The discrepancies were partly due to offsets in the wind direction and/or the overestimation of wind speed. Based on these limited data WRF/Chem performed for the Fairbanks metropolitan area acceptably. Unfortunately, with daily data from just one site and 1-in-3-days data from just three sites (of which one site coincided with the site of daily data) a more thorough analysis is impossible.

The episode was relatively cloudy (e.g. Figure 4) and clouds attenuated the lidar signals. Thus, only few cases existed where the lidar reached the smoke in the ABL. The qualitative comparison with these CALIPSO lidar curtains suggested that WRF/Chem acceptably simulated the locations of particulate matter in the ABL. On some days, offsets occurred in the horizontal position of the smoke plumes. These discrepancies can be explained by the timelag between the WRF/Chem cross-sections and the overestimation of wind-speeds. The offsets were of similar magnitude as those found by [47] in another subarctic study.

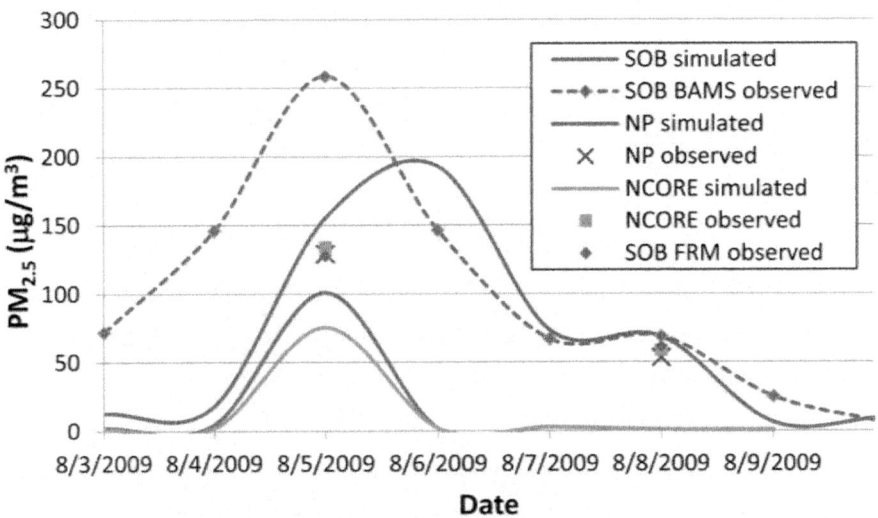

Figure 3. Temporal evolution of hourly $PM_{2.5}$ concentrations as simulated (solid lines) and observed (markers) at the State Office building in downtown Fairbanks (SOB) with the BAMS and FRM instruments, in the city of North Pole (NP), and at the proposed NCORE site in Fairbanks. North Pole belongs to the Fairbanks metropolitan area and is located about 21 km southeast of Fairbanks. August 3 to 4 were during the spin-up of the chemical fields.

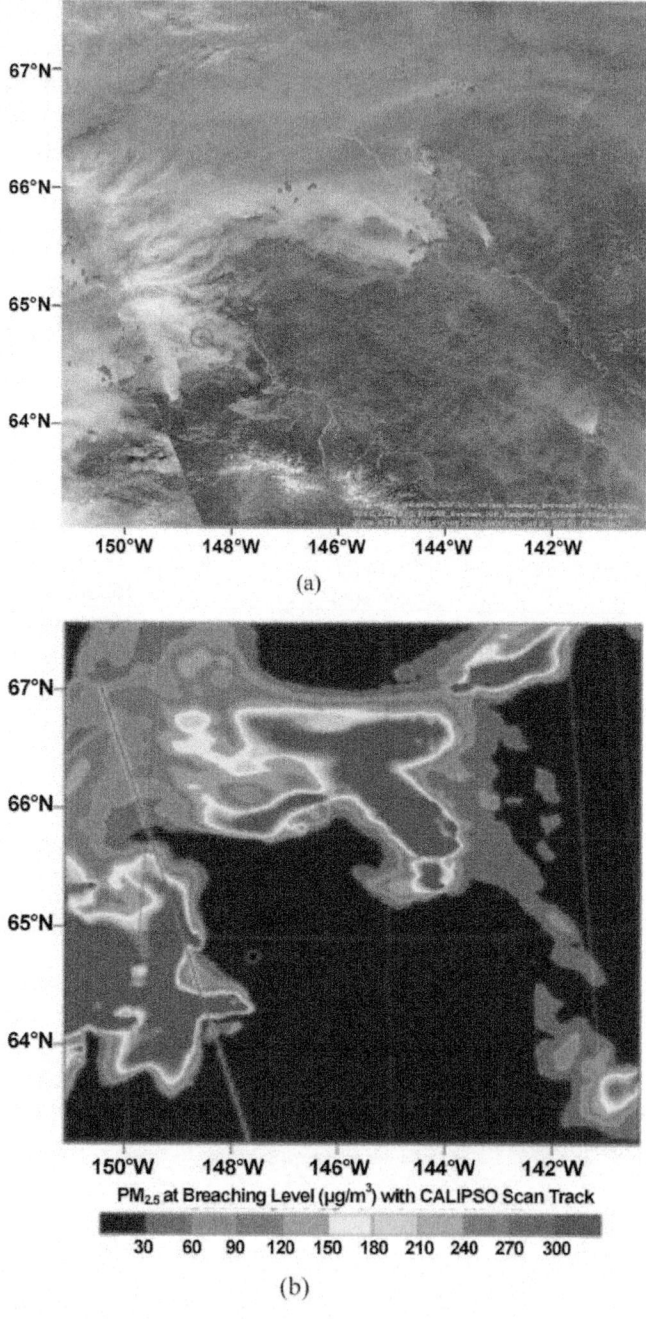

Figure 4. (a) MODIS visible channel data at the time of the CALIPSO path (red line in part (b)) at 1201 UTC, and (b) WRF/Chem simulated PM$_{2.5}$ distribution at 1200 on

August 5, 2009. The red dots and blue circle in (a) mark the locations of active fires and the AERONET site at Bonanza Creek (30 km southwest of Fairbanks). The black dot in (b) marks the location of Fairbanks.

Comparison of WRF/Chem vertical-integrated horizontal distributions of smoke extend with MODIS data (e.g. Figure 4) suggested similar spatial offsets of the WRF/Chem-simulated smoke plume as found in the comparison with CALIPSO data.

Comparison of simulated and observed cloud distributions revealed that in mountainous terrain, WRF/Chem underestimated convection related to slope winds. This shortcoming was because WRF/Chem used the mean terrain height as representative for the terrain height within each grid-cell. Consequently, steep or small valleys were of subgrid-scale. Furthermore, WRF/Chem had difficulties capturing some of the cirrus seen in MODIS and/or CALIPSO data due to the coarse vertical grid resolution at these heights.

Above high-level clouds and in the upper troposphere, simulated PM_{10} concentrations showed a homogeneous distribution with marginal changes over time. On the contrary, the CALIPSO data suggested more heterogenous distributions. This discrepancy was due to the coarse resolution of WRF/Chem in the upper troposphere. For the same reason, WRF/Chem failed to simulate some of the cirrus clouds and the full vertical extent of high reaching convection (e.g. Figure 5).

General Findings Regarding Sampling by UAV

The virtual sampling focused on the 60 km ′ 60 km area centered over the Crazy Mountain fires complex (Figure 1). Following [3] 20 h mean values determined from virtual sampling were compared to the 20 h means of the WRF/Chem data that served as the "grand truth" dataset.

The 20 h mean distributions from sampling at three altitudes were able to capture the vertical gradients for air and dewpoint temperature (O_3) that naturally increase (decrease) with height in the ABL. These distributions also captured pertubations of these general features when the perturbations occurred due to advection.

After the cold front passed on August 6, concentrations of particulate matter were quasi-uniform at all heights. Thereafter, heterogeneity increased as time progressed.

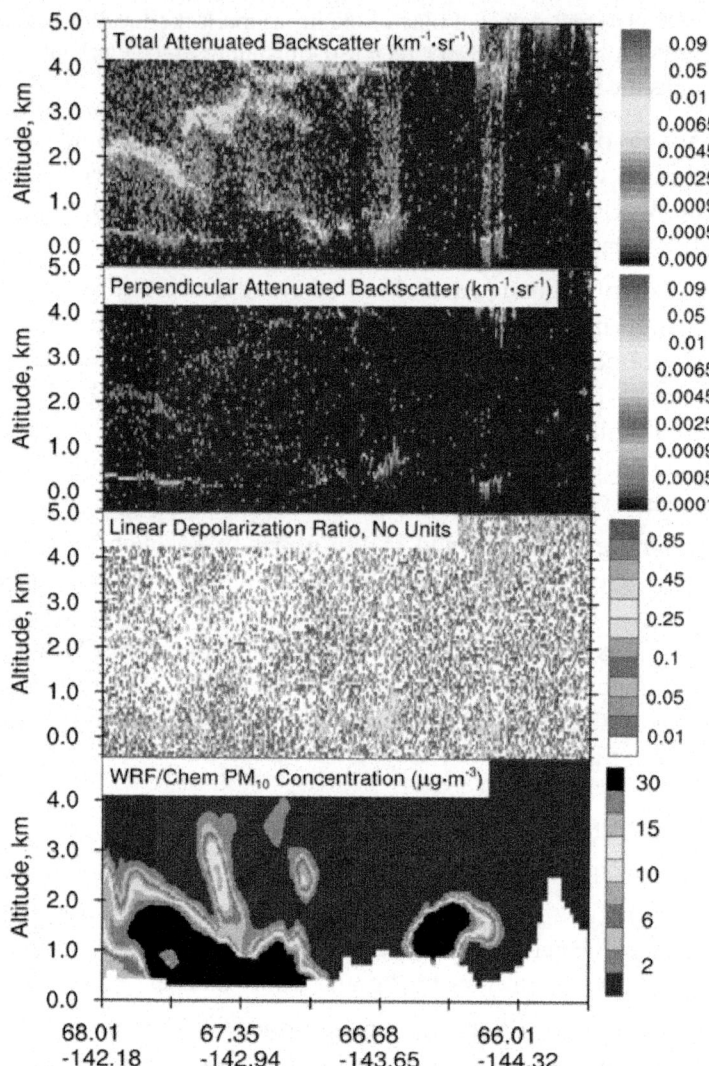

Figure 5. Example of evaluation by CALIPSO data: Nighttime total attenuated back-scatter, perpendicular attenuated backscatter, and LDR for the CALIPSO path on August 5, 2009 at 1201 UTC. Note the increase in backscatter from the surface up to 2 km height. An increase in depolarization is visible at the surface up to 2 km above ground. These data indicate particulate matter extending from the surface upward. WRF/Chem simulated total liquid and solid mixing ratio distribution (contours) and PM_{10} concentrations (color shades) at 1200 UTC.

The following applied to all sampling heights, flight patterns, and flight speeds:

- For all constituents, 20 h mean concentration distributions differed stronger from the "grand truth" on the day with the frontal passage than on the days prior to or after the event.

- The 20 h mean distributions from sampling differed strongest among each other and from the "grand truth" for field quantities with a distinct diurnal course.

- The virtual UAV may sample in areas of extreme values. However, the 20 h means smoothed the distributions due to changes in wind directions and because the magnitudes of minima and maxima varied in time.

- The likelihood for sampling in the region of maximum values decreased as the spatial-temporal varibility in plume location increased.

- Correlation between sampled and "grand truth" means decreased, and errors (e.g. RMSE, NMB, FB) increased with increasing natural spatio-temporal heterogeneity of the field quantity.

Temperature and Dewpoint Temperature

Our discussion focused on an area that a Scan Eagle could cover within 20 h of sampling (Figure 1). For each day of August 3 to 10, we determined the distribution of 20 h means from the WRF/Chem data over the sampling domain as the reference of the "grand truth". Sampling at three heights with three different speeds and flight patterns yielded the following.

On average (August 3 to 10), temperatures differed about $1°C·100$ m^{-1} between the three flight levels, i.e. the lower ABL was nearly dry-adiabatic except for August 6, the day of the cold front passage. The virtual sampling reflected this vertical behavior well in all cases.

Typically, distributions derived from sampled temperatures differed the strongest from the "grand truth" where the virtual UAV sampled at times around the daily maximum temperature. Sampled and "grand truth" mean temperatures agreed best on days without frontal activity in the sampling domain. On these days, the design of the flight patterns barely played a role for the differences between sampling-derived and "grand truth" distributions of 20 h mean temperatures.

Discrepancies between sampling-derived and "grand truth" distributions of air temperatures at low altitude exceeded those at high altitude in the ABL (Figure 6). At 200 m, surface heterogeneity still affected vertical motions and sensible heat flux densities that documented themselves in notable horizontal variations of temperature. At the top of the ABL, however, the fluxes blended to a relative homogeneous distribution. The diurnal temperature course vanished at this height. Note that RMSEs and SDEs from heterogenous samples are naturally larger than from comparatively more homogeneous samples [64].

On days without any changes in the synoptic conditions, mean differences between the temperature distributions obtained for the three flight patterns and the "grand truth" were less than 2°C. However, local differences reached up to 7°C. Typically, at the same height, sampled distributions of air temperatures differed least among each other in the middle, and largest along the boundaries of the sampling domain (Figure 6). This behavior was because the virtual UAV sampled the middle part of the sampling domain at about the same time, but the boundaries at quite different times when flying the three different patterns (cf. Figure 1).

On August 6, the passage of a cold front led to huge discrepancies between sampling-derived distributions in areas affected by the front during the 20 h. Since the cold front sloped backward with height, notable discrepancies occurred over a smaller region at 1000 m height than at 500 m or 200 m (Figure 6). Investigation showed that during frontal events, traveling north-south or vice versa with only 4 km legs for turning yielded distributions of mean temperatures with 15% positive (sampled vs. "grand truth") normalized mean biases and fractional mean biases in regions where the front moved in. Long west-east/east-west sampling legs after the passing of the cold front led to significant (p-value < 0.0001) negative NMB (−34%) and FB (−28%). The default sampling pattern, however, yielded an acceptable mean distribution as the positive and negative biases from sampling prior to and after the frontal passage cancelled each other out.

The results from the default and third sampling patterns only slightly differed as both sampled the front at similar times in about the same location. Areas of small and large differences were rotated among the default and third patterns reflecting the relationship of the two sampling patterns (Figure 1, Figure 6). However, the second sampling pattern sampled the moving cold front at a quite different time and location than the default and third patterns. Thus, the distribution of 20 h mean temperatures obtained from sampling showed considerable differences from the "grand truth." Based on this finding, one has to conclude that under conditions of frontal activity, flight patterns can affect the mean distributions derived therefrom.

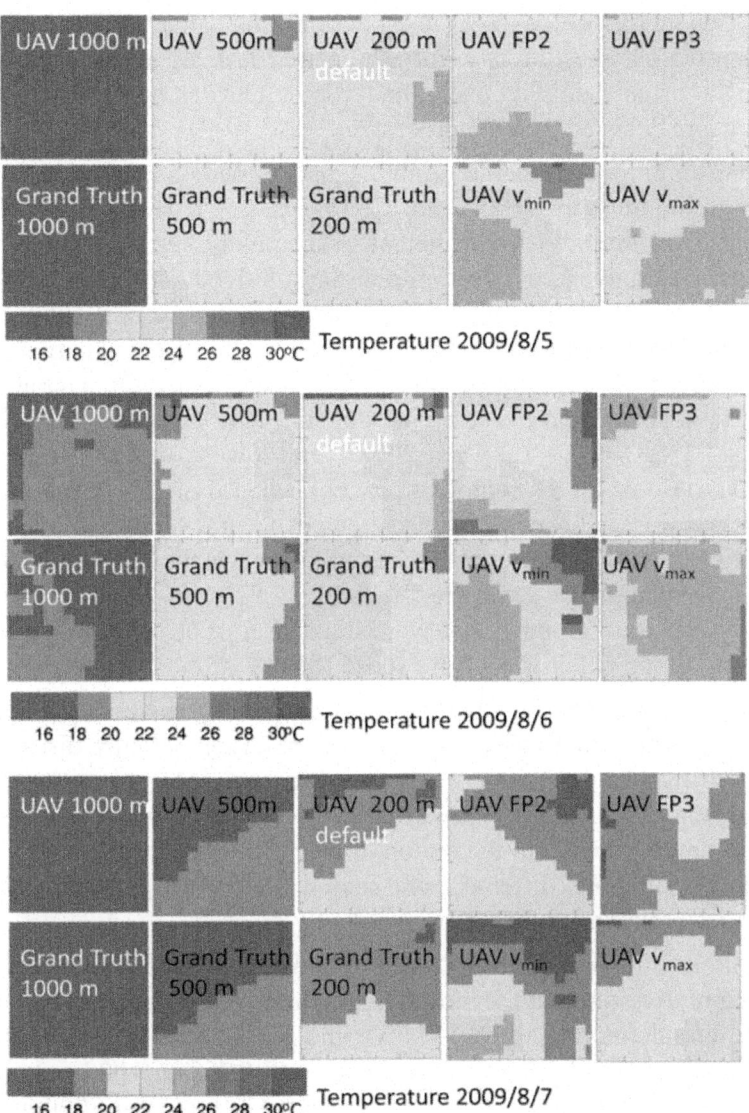

Figure 6. Distributions of mean temperatures as obtained (top left to bottom right) from virtual sampling at 1000 m, 500 m, 200 m (default), flight pattern 2 (FP2) at 200 m, flight pattern 3 (FP3) at 200 m, from the "grand truth" at 1000 m, 500 m, and 200 m, from sampling at stall speed (v_{min}) at 200 m, and maximum speed (v_{max}) at 200 m for prefrontal conditions on August 5 (first two rows of panel), conditions during the frontal passage on August 6 (middle two rows of panel), and after the front passed on August 7 (last two rows of panel). Panels show 20 h means for all sampling designs except for those using stall and maximum speed.

Consequently, one has to choose the sampling pattern that is most suitable for the research question/task for which the data are needed. This means, applying a virtual sampling using the forecast data can be of help in the flight planning and decision-making process.

According to the "grand truth", episode-mean temperatures over the sampling domain were 20°C ± 2.6°C, 19.7°C ± 2.7°C, and 13.9°C ± 2.5°C at 200 m, 500 m and 1000 m height, respectively. On this spatio-temporal mean, the best agreement between sampling-derived and "grand truth" mean temperatures according to the RMSE and correlation were flight pattern 2 at cruise-speed, flight pattern 1 at maximum speed, and flight pattern 3 at cruise-speed at 200 m, 500 m and 1000 m height, respectively. On average over all days of the episode and the sampling domain, RMSE, and SDE were highest (3.2°C - 3.5°C, 1.6°C - 1.9°C) at 200 m, and lowest (0.6°C - 0.7°C, 0.7°C - 0.8°C) at 1000 m for all sampling patterns when flying at cruise speed. The mean biases were of the order of measurement accuracy independent of altitude and sampling patterns at cruise speed. Except for the day with the frontal passage, correlations between the distribution of 20 h mean temperatures from sampling and the "grand truth" were highest at 1000 m (0.8 - 0.9) for the various sampling patterns at cruise speed [8].

The investigations on flight speeds suggested that the signature of the diurnal cycle became more obvious in the mean temperature distributions at stall speed than at maximum speed (e.g. Figure 6). During frontal passages, however, the virtual UAV's speed became decisive for whether the sampling occurred predominantly in the region in front of or behind the cold front. The sampled distributions differed from each other for all patterns with largest differences occurring at 200 m.

Looking at the various flight patterns and cruising speeds revealed that the sampling underestimated the mean temperatures on average over all days and the sampling domain by 0.9°C - 2.6°C, and 1.5°C - 2°C at 200 m and 500 m height, respectively, but overestimated it up to 0.4°C at 1000 m height. The sampling suggested about twice as high spatial variation at all altitudes than was present in the "grand truth." Overall, the above findings suggested that determining area-temporal mean values for an area of 60 km ´ 60 km, about the size of high- resolution climate models, provided similar uncertainty than deriving them for small areas, i.e. in our case 4 km ´ 4 km areas. However, providing area means for large areas (e.g. the entire Interior Alaska) would require flying several UAVs in adjacent areas at the same time.

In Interior Alaska summer, moist flux densities are not large, for which dewpoint temperatures do not change quickly except when a front moves in.

Dewpoint temperatures showed little diurnal variability in the sampled values and the "grand truth" (therefore not shown).

In the U.S. standard atmosphere, dewpoint temperature decreases at a rate of $0.172°C \cdot 100 \text{ m}^{-1}$[65]. Consequently, distributions of mean dewpoint temperatures sampled at different altitudes have to differ [8]. Wind shear and advection changed the dewpoint-temperature profiles. Thus, dewpoint temperatures decreased at non- constant rate with height. Consequently, dewpoint-temperature distributions from virtual sampling at different heights differed from each other between 1°C and 3°C. The latter occurred on August 5 due to the approaching low-pressure system.

On average over the episode and sampling domain, "grand truth" dewpoint temperatures were $1°C \pm 0.3°C$, $0.5°C \pm 1.5°C$, and $-0.3°C \pm 1.6°C$ at 200 m, 500 m, and 1000 m, respectively. Over the episode and sampling domain, spatio-temporal variability obtained from sampling dewpoint temperatures at 200 m exceeded that of the "grand truth" by threefold for all three sampling patterns. However, the episode sampling domain mean temperature was captured independent of the flight pattern. In contrast to the 200 m level, sampling well captured the spatio-temporal variability of dewpoint temperatures at 500 m and 1000 m height. The virtual UAV's speed had marginal impact on the differences between distributions of mean dewpoint temperatures from sampling and those derived from the "grand truth" except for August 6 when the cold front went through.

On 6 August, sampling at stall speed showed increases in dewpoint temperatures up to 4°C in the western part of the sampling domain. At cruise and maximum speeds, most of the virtual sampling occurred in front of the cold front where dewpoint temperatures were still low.

On average over the episode and sampling domain, discrepancies between sampled and "grand truth" temperatures decreased with height. Sampled mean dewpoint temperature distributions and the "grand truth" correlated the least for sampling at stall speed as part of the area was not sampled within the 20 h flight duration.

Inert Gases

Even though CO is part of a series of chemical reactions that form photochemical smog, its mean atmospheric lifetime is about 60 days [65] [66]. Thus, CO commonly serves as a tracer of wildfire smoke. By adopting this strategy, we took CO as a representative for an inert gas. Of course, WRF/Chem considered reactions of CO along with aldehydes as part of photochemical smog formation (cf. [26]).

During a wildfire, CO concentrations increase downwind of the fire due to transport. The NAAQS for CO is 35 ppm on 1 h and 9 ppm on 8 h average [14]. In Interior Alaska, typical ambient mean CO concentrations are about 3 ppm [8].

For each day of August 5 to 10, we determined the distribution of 20 h means from the WRF/Chem data over the sampling domain as the reference ("grand truth"). August 3 and 4 were discarded from the analysis to permit the chemical fields to spinup. This procedure was applied for all chemical species and particulate matter as well.

Prior to the frontal passage, locally, CO concentrations exceeded the 8 h average in the sampling domain [8]. Vertical differences were strongest prior to the frontal passage, as CO had built up to a smoke plume on the days before. On August 5, for instance, differences of up to 13.18 ppm occurred between 200 m and 1000 m. On the day of the frontal passage, according to the "grand truth", highest CO concentrations existed in the western part of the sampling domain and the southeastern corner at all three heights (Figure 7). After the frontal passage, CO reached clean air background concentrations. After August 6, CO concentrations built up again as time progressed [8].

In the sampling domain, CO concentrations decreased about 13% at most between 200 m and 1000 m in both the distributions from virtual sampling and the "grand truth" (Figure 7). The 20 h mean CO concentrations differed between these heights due to the proximity to the wildfire. On average over the episode (August 5-10), absolute differences between the CO concentrations at 200 m and 1000 m locally reached 2 ppm, i.e. a strong vertical gradient of CO existed in the ABL due to the wildfire. The virtual sampling well captured that vertical differences in CO concentrations were least on the day after the frontal passage (not shown).

At 200 m, for the first and second flight patterns, the 20 h mean CO distributions based on virtual sampling showed larger spatio-temporal variability than the "grand truth". On the contrary, using the third flight pattern underestimated the spatio-temporal variability on average over the period and sampling domain.

At 500 m height, the second flight pattern suggested twice as high spatio-temporal variability than the "grand truth." The default and third flight patterns showed the same spatio-temporal variability as the "grand truth." This finding differs from that of temperature and dewpoint temperature due to the stronger spatial (horizontal and vertical) heterogeneity of CO. On average over the episode, sampling at 500 m represented the distribution of relatively higher and relatively lower CO concentrations the best because the wind field was less turbulent at this height than at 200 m. Compared to CO concentrations at the 200 m height, the differences between high and low concentrations were smaller at 500 m height.

Since the height of the ABL varied during the 20 h, virtual sampling at 1000 m was sometimes above the inversion where concentrations were lower than below the top of the ABL. At 1000 m, all sampling patterns underestimated the spatio-temporal variability by a factor of two. Air-quality models are considered to have high preformance when predicted concentrations agree within a factor of two with the observations [55] -[67].

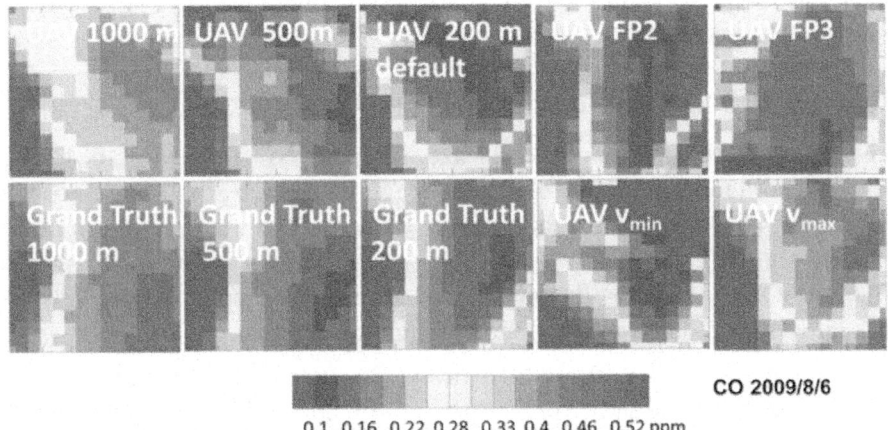

CO 2009/8/6

0.1 0.16 0.22 0.28 0.33 0.4 0.46 0.52 ppm

Figure 7. Distributions of mean CO concentrations as obtained (top left to bottom right) from virtual sampling at 1000 m, 500 m, 200 m (default), flight pattern 2 (FP2) at 200 m, flight pattern 3 (FP3) at 200 m, from the "grand truth" at 1000 m, 500 m, and 200 m, from sampling at stall speed (v_{min}) at 200 m, and maximum speed (v_{max}) at 200 m during the frontal passage on August 6. Panels show 20 h means for all sampling designs except for those using stall and maximum speed. Note that discrepancies between the distributions from virtual sampling and the "grand truth" on other days look similar (therefore not shown).

When applying the same quality criterion for good agreement to the 20 h mean CO distributions derived from sampling vs. those of the "grand truth" like in air-quality modeling our findings mean that the calculation of 20 h mean distributions from UAV data will provide valuable results.

These findings for the sensitivity of derived CO distributions to flight patterns differed from that of temperature and dewpoint temperature. Recall for both air and dewpoint temperatures, the first and third patterns captured the 20 h mean distributions of the "grand truth" in a similar way. Obviously, which sampling pattern is the most suitable depends on the vertical profiles and horizontal distributions of the sampled quantities. These distributions were quite different for CO and air temperature/dewpoint temperature.

The injection height for all wildfire-released species was calculated inline by WRF/Chem. Injection height varied with time reaching up to 4 km above ground level at some times. This means species and temperature distributions were not collocated in space and time.

These differences in the distributions of the sampled quantities suggest that a numerical forecast and virtual sampling of the forecasted quantities may be needed to decide on flight levels and sampling patterns for the various quantities to be observed by a cohort of UAVs. In other words, our findings mean that different flight patterns are to be considered for UAVs depending on the mounted instrument.

Reactive Primary Pollutants

Among other things, wildfires release SO_2 and NO [22] [66] [68]. These primary pollutants undergo chemical reactions thereby building secondary pollutants. They are also precursor gases for aerosol formation [66]. In Interior Alaska, annual mean near-surface SO_2 concentrations are about 35 ppb [8]. The NAAQS for SO_2 for 1 h and 3 h are 75 ppb and 0.5 ppm, respectively [14]. The latter is not to be exceeded more than once per year.

Prior to the frontal passage, 20 h mean SO_2 concentrations ranged between 2 ppb outside the plume and 18 ppb in the plume at 200 m (Figure 8). The relatively low 20 h SO_2 concentration means resulted from the plume's meandering due to the calm winds. At 500 m, 20 h mean maximum concentrations exceeded 4 ppb. Local inversions hindered vertical exchange in some areas of the sampling domain. Thus, at 1000 m, 20 h mean SO_2 concentrations were elevated or represented clean air background concentrations. Immediately after the frontal passage, 20 h mean SO_2 concentrations corresponded to clean air background values at all three heights (<1 ppb). On the day after the frontal passage, highest SO_2 concentrations occurred in the western part of the sampling domain at 1000 m height building up over time due to advection of wildfire smoke. Due to the weak winds peaks in SO_2 concentrations occurred in similar locations at the three altitudes with slightly higher concentrations at increasing altitude.

The 20 h mean distributions of SO_2 constructed from sampling showed locally positive and negative biases as compared to the "grand truth." Investigations showed that SO_2 concentrations decreased at onset of twilight (~0004 Alaska Daylight Time (AKDT = UTC − 8 h)) hinting at photolytic reactions being involved. During the episode of this study, sunrise occurred between 0430 and 0500 AKDT and sunset was between 2300 and 2230 AKDT. The decrease in SO_2 and sulfate particulate matter showed no correlation [8]. Thus, gaseous SO_2 forming solid aerosols can be excluded as cause of the decrease.

The virtual UAV collected data for 20 h. Hence, it took samples at different times of the diurnal course of SO_2 concentrations. In areas where the virtual UAV sampled when concentrations were low in the diurnal course, the 20 h means derived therefrom underestimated the "grand truth" 20 h mean SO_2 concentrations (cf. Figure 1, Figure 8). The opposite was true for areas where sampling occurred only when concentrations were at their highest in the diurnal course. Consequently, the different sampling patterns yielded different answers for the distributions of 20 h mean SO_2 concentrations.

The cold front reset the SO_2 concentrations to clean air background concentrations. Thus, on August 7, the distributions of 20 h mean SO_2 concentrations from sampling and the "grand truth" agreed well independent of the flight patterns in most of the sampling domain except for its southeast corner (Figure 8). During this time, a northwest wind pushed the smoke plume from the wildfire into this area. Differences were due to the time of sampling in this area of the redeveloping smoke plume.

On August 8, wind direction shifted, for which the highest SO_2 concentrations occurred farther north and in the center of the sampling domain (not shown). Due to the calming of the winds, the plume dispersed more strongly than the day before. Nevertheless, on all days, the largest differences between the distribution of 20 h means from SO_2 sampling and the 20 h "grand truth" means occurred along the corners of the sampling domain for all flight patterns (e.g. Figure 8).

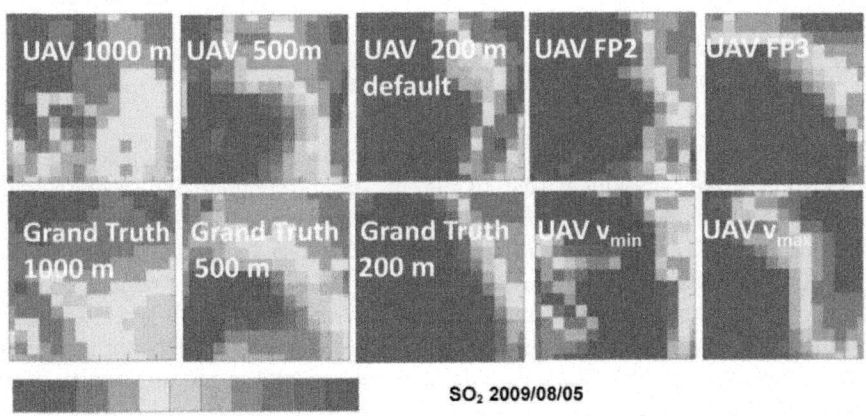

SO₂ 2009/08/05

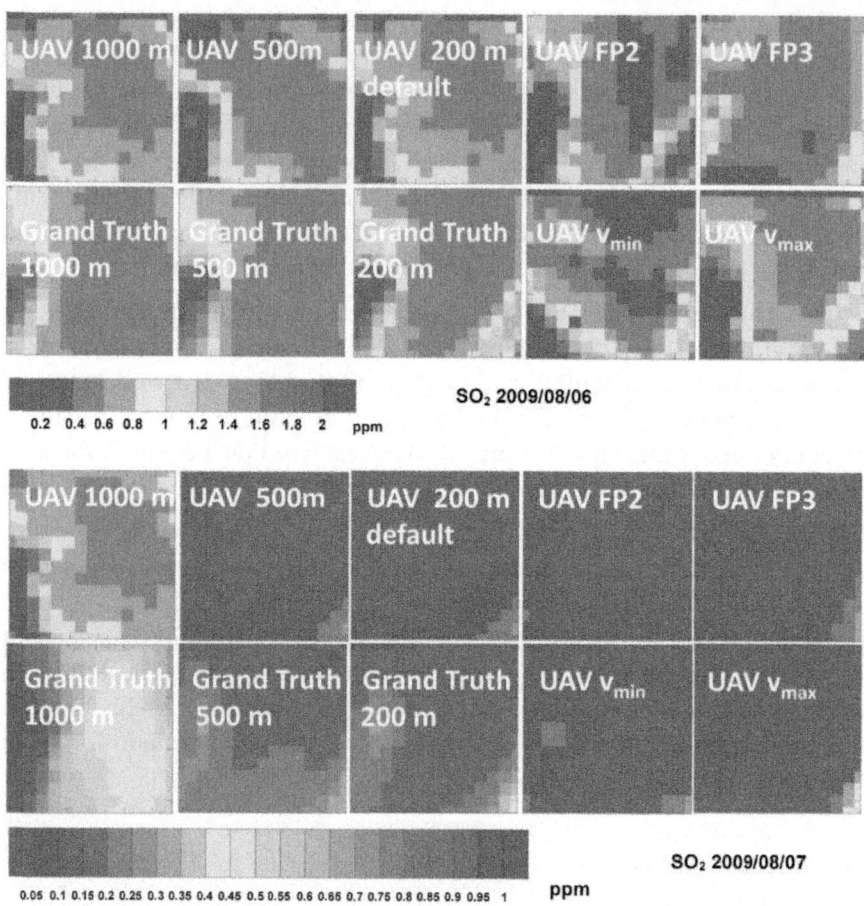

Figure 8. Distributions of mean SO$_2$ concentrations as obtained (top left to bottom right) from virtual sampling at 1000 m, 500 m, 200 m (default), flight pattern 2 (FP2) at 200 m, flight pattern 3 (FP3) at 200 m, from the "grand truth" at 1000 m, 500 m, and 200 m, from sampling at stall speed (v_{min}) at 200 m, and maximum speed (v_{max}) at 200 m for prefrontal conditions on August 5 (first two rows of panel), conditions during the frontal passage on August 6 (middle two rows of panel) and after the front passed on August 7 (last two rows of panel). Panels show 20 h means for all sampling designs except for those using stall and maximum speed. Legend for August 7 differs from those of August 5 and 6.

Here, data were sampled either within very short time or at long time apart. Thus, one may conclude that both short and long temporal increments between sampling can cause strong biases from the 20 h means of the "grand truth" for species that have a notable diurnal course (cf. Sections 3.1, 3.3, Figure 1).

Typically, independent of the virtual UAV's speed, 20 h mean SO_2 concentrations were overestimated and underestimated within areas of high and low concentrations, respectively (e.g.Figure 8). The 20 h mean SO_2 concentrations derived from virtual sampling at different speeds roughly agreed with each other and the "grand truth." However, at stall speed, differences throughout the sampling domain were larger and less localized than at cruising or maximum speeds. When sampled at maximum speed, 20 h mean SO_2 concentrations from sampling correlated the strongest with those of the "grand truth" (0.7 at 200 m, 0.9 at both 500 m and 1000 m).

However, at 200 m altitude, 20 h SO_2 concentration means from sampling captured the spatio-temporal variability best when flying at stall speed. Flying at maximum speed suggested 50% higher spatio-temporal variability than existed according to the "grand truth" at 200 m. On average, at 500 m altitude, spatio-temporal variability was underestimated at all speeds by 44% to 77%. The strongest (least) underestimation occurred at cruising (stall) speed. At 1000 m, on average, the largest underestimation of spatio-temporal variability occurred at stall speed, while flying at maximum speed provided the best results and highest correlation of sampled and "grand truth" 20 h mean SO_2 concentrations. Based on these findings, one has to conclude that sampling SO_2 concentrations at high speeds minimizes errors. This finding is because the virtual UAV needs less time to cover the entire sampling domain. Thus, signals of extremes in the diurnal course have less impact at highest than at slower speeds.

Due to the sparse population and synoptic situation, the main source of NO in the sampling domain was the Crazy Mountain fires. Due to the reactivity of NO, the NAAQS considers NO_2 with a 1 h average of 100 ppb [14].

Like for SO_2, NO has a diurnal cycle due to photochemical reactions [66]. The analysis of the results for NO confirmed those of SO_2. Concentrations were overestimated (underestimated) in areas of high (low) NO concentrations (Figure 9). Like for SO_2, NO concentrations reset to clean air background values when the cold front passed the sampling domain, and the smoke plume re-developed after the front had passed.

On episode (August 5 to 10) and sampling domain average, sampled and "grand truth" NO concentrations agreed best with respect to the combined RMSE and correlations scores for flight pattern 1 at maximum cruise speed at all heights.

In summary, the virtual sampling showed that due to the diurnal cycle of pollutants involved in photochemistry 20 h mean distributions locally fail to capture the 20 h mean of the "grand truth" undoubtly. For all days after spinup, i.e. also on the days without frontal passage, the obtained distributions were sensitive to when the UAV passed an area. Analysis suggested that data should be separated for daylight and dark hours to determine daylight and nighttime mean distributions instead of 20 h mean distributions [8].

Secondary Pollutants

Secondary pollutants like O_3 form by reactions involving primary pollutants. The NAAQS for O_3 is an 8 h average concentration of 75 ppb [14]. In Interior Alaska, typical O_3 concentrations are about 40 ppb [8].

Recall that during the episode of our study, complete darkness occurred only for about 4 to 5 hours. According to the WRF/Chem data, O_3 concentrations showed no distinct minimum during daylight despite the reactions with NO and VOCs. The distributions of 20 h mean O_3 concentrations showed an increase of O_3 with increasing height. Typically, 20 h mean O_3 concentrations ranged between 36 and 44 ppb, 40 and 50 ppb, and 42 and 52 ppb at 200 m, 500 m and 1000 m height, respectively. Overall, the O_3 distributions showed low spatial features at the three heights and on all days. Comparison of the O_3 concentrations prior to and after the cold front indicated some ozone formation due to the wildfire emissions. The low spatial and temporal changes in O_3 concentrations yielded for the default and second flight patterns provided broadly similar distributions of 20 h means (Figure 10). However, these distributions failed to capture that of the "grand truth." The third flight pattern captured the location of relatively high and low concentrations best. Nevertheless, discrepancies between 20 h means from virtual sampling and the "grand truth" were about ±4 ppb for all three flight patterns. Obviously, sampling of O_3 concentrations is sensitive to the UAV's speed (cf. Figure 10). Typically, 20 h mean O_3 concentrations were higher for flying at stall or maximum speed as compared to flying at cruising speed. Distributions of 20 h mean O_3 concentrations agreed best with those of the "grand truth" when the virtual

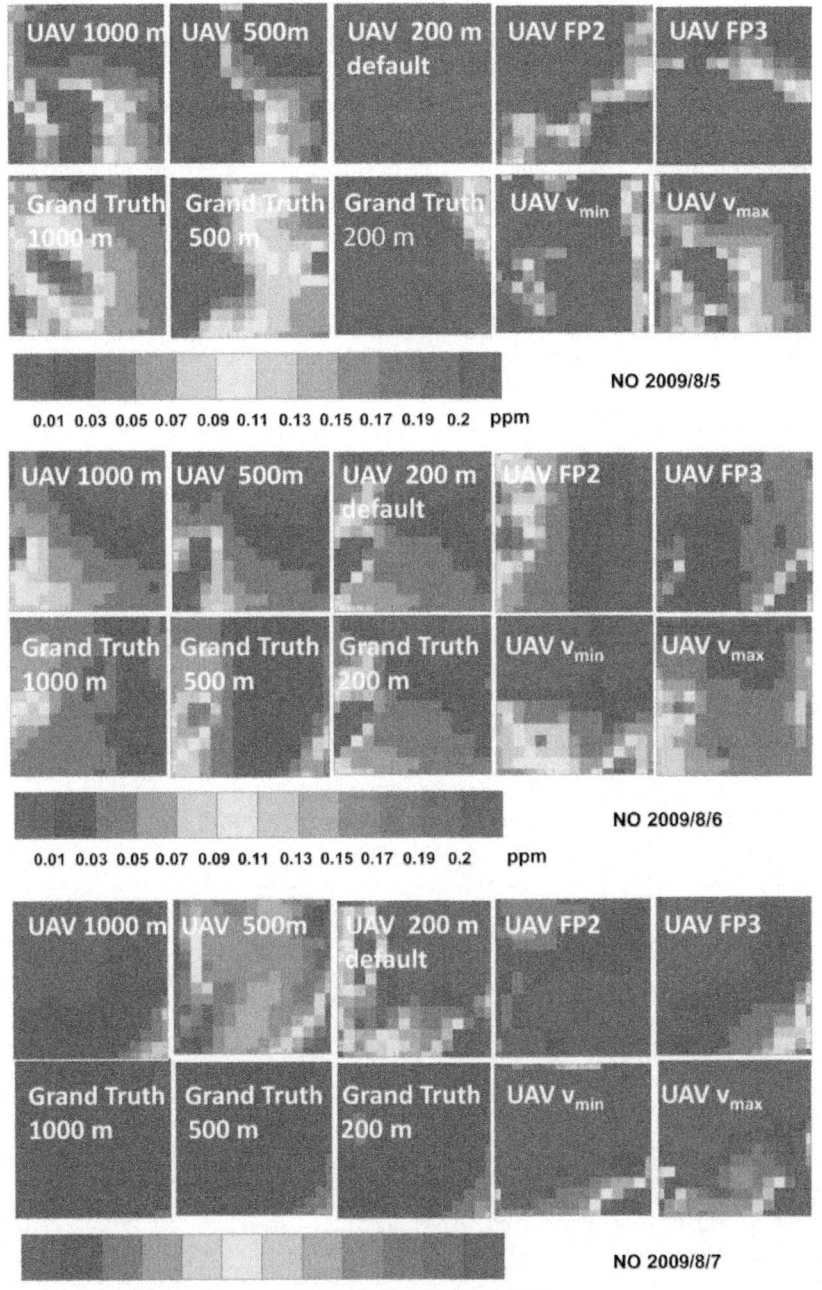

Figure 9. Distributions of mean NO concentrations as obtained (top left to bottom right) from virtual sampling at 1000 m, 500 m, 200 m (default), flight pattern 2 (FP2) at 200 m, flight pattern 3 (FP3) at 200 m, from the "grand truth" at 1000 m, 500 m, and

200 m, from sampling at stall speed (v_{min}) at 200 m, and maximum speed (v_{max}) at 200 m for prefrontal conditions on August 5 (first two rows of panel), conditions during the frontal passage on August 6 (middle two rows of panel) and after the front passed on August 7 (last two rows of panel). Panels show 20 h means for all sampling designs except for those using stall and maximum speed. UAV flew at cruise speed. Since the obtained 20 h mean O_3 distributions were best at cruise speed, an optimum sampling speed may exist that could be determined by virtual sampling.

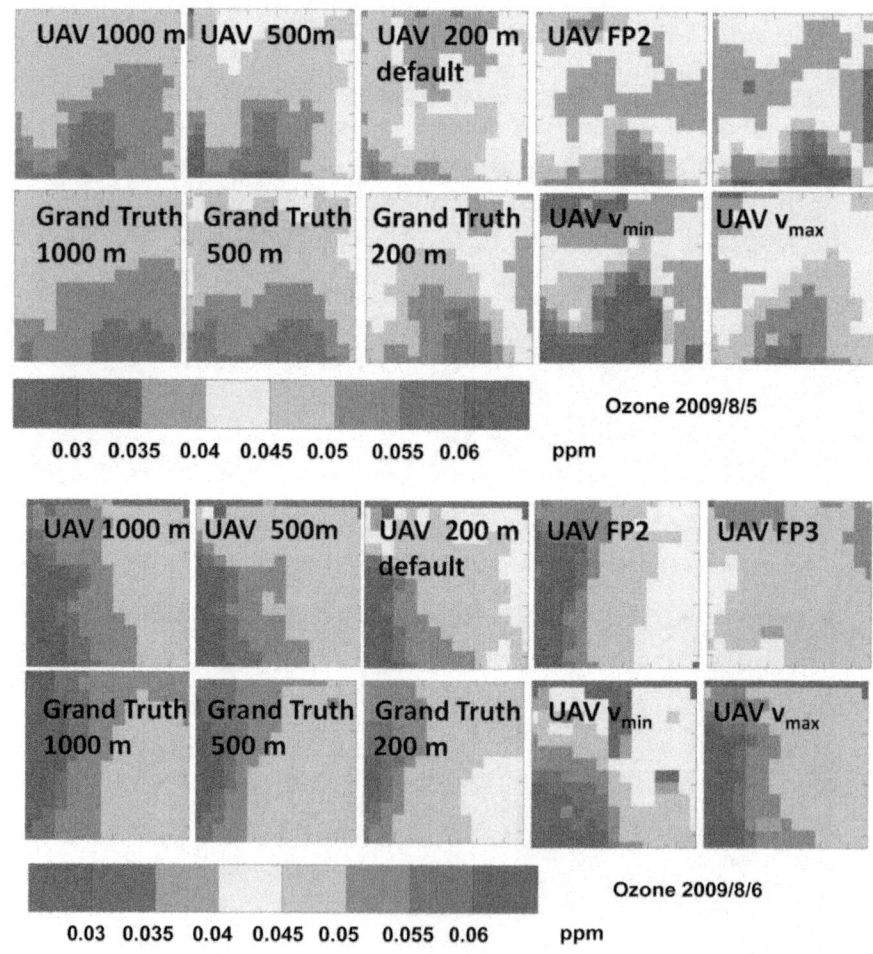

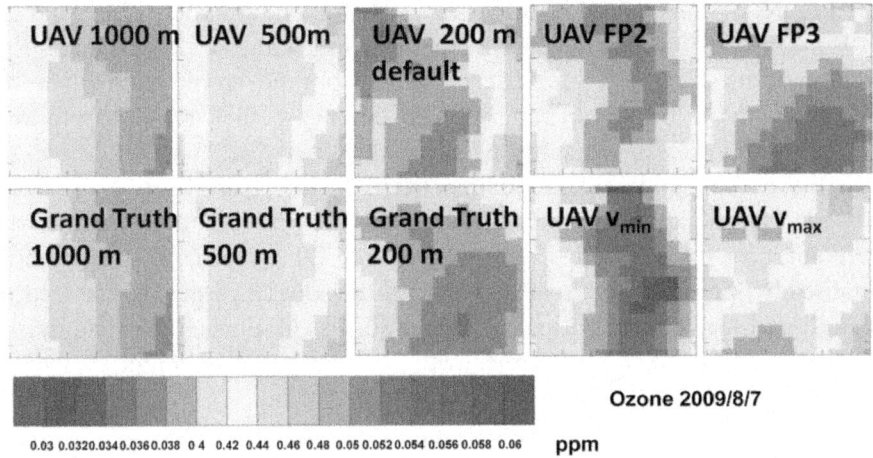

Figure 10. Distributions of mean O_3 concentrations as obtained (top left to bottom right) from virtual sampling at 1000 m, 500 m, 200 m (default), flight pattern 2 (FP2) at 200 m, flight pattern 3 (FP3) at 200 m, from the "grand truth" at 1000 m, 500 m, and 200 m, from sampling at stall speed (v_{min}) at 200 m, and maximum speed (v_{max}) at 200 m for prefrontal conditions on August 5 (first two rows of panel), conditions during the frontal passage on August 6 (middle two rows of panel) and after the front passed on August 7 (last two rows of panel). Panels show 20 h means for all sampling designs except for those using stall and maximum speed. Legend for August 7 differs from those for August 5 and 6.

Particulate Matter

Particulate matter (PM) can form in the atmosphere from precursor gases by gas-to-particle conversion [16] [66]. The wildfires were the main source for particulate matter in this area of the model domain. Currently, the NAAQS for 24 h mean PM_{10} and $PM_{2.5}$ are 150 and 35 $\mu g \cdot m^{-3}$, respectively [14]. Except for the Fairbanks metropolitan area and in the downwind of wildfires, typical concentrations are below 2 $\mu g \cdot m^{-3}$ [29].

According to the "grand truth" during the episode, 20 h mean PM_{10} concentrations were highest at the 200 m flight level except for August 6 (Figure 11). Then 20 h mean PM_{10} concentrations reached up to 71.9 $\mu g \cdot m^{-3}$ and 6.23 $\mu g \cdot m^{-3}$ at 1000 m and 200 m, respectively. At the lower flight levels, the UAV sampled in the cold sector. Here precipitation had removed already PM_{10}. At 1000 m, the virtual UAV sampled data in the polluted air of the warm sector.

The obtained 20 h mean distributions of PM_{10} concentrations depended much more on the flight patterns than the meteorological or gaseous quantities (cf. Figure 6 to Figure 11), especially during precipitation. On August 6, 20 h mean distributions derived from sampling with the default and second flight patterns showed greater discrepancies from the "grand truth" than those obtained with the third flight pattern. In the latter case, the virtual UAV followed the precipitation. Consequently, it sampled lower PM_{10} concentrations than with the default and second flight patterns. Therefore, absolute differences between 20 h mean PM_{10} concentrations from sampling and the "grand truth" were smallest for the third flight pattern. The 20 h mean PM_{10} distributions obtained by the three flight patterns even differed with respect to the locations of high and low concentrations. In general, distributions from sampling and the "grand truth" correlated higher with increasing cruise height, as there the distributions were more homogeneous than at 200 m.

Sampling at maximum speed permitted capturing high concentrations better than sampling at the other speeds (Figure 11). Unlike for different flight patterns, the differences between the 20 h mean distributions from sampled data and the "grand truth" showed similar spatial structures. This means that sampling underestimated the 20 h mean PM_{10} concentrations in the same area independent of flight speed. Speed determined the magnitude of underestimation.

According to the WRF/Chem data, like for PM_{10}, the 20 h mean horizontal distributions of $PM_{2.5}$ varied typically the strongest at the 200 m flight level (Figure 11, Figure 12). The reason was that the sources were at the ground not far from each other and the landscape modified the flow. As the smog-plume ascended, it experienced mixing. At the top of the ABL, pollutants accumulated leading to high concentrations and horizontal variations at a larger scale than at 200 m height. On average, $PM_{2.5}$ concentrations increased with increasing height in the ABL. Consequently, at 200 m, 20 h mean minimum $PM_{2.5}$ concentrations were lower than the minimum $PM_{2.5}$ concentrations at 500 m and 1000 m. On August 8, for instance, local maximum differences between the 20 h mean concentrations at 200 m and 1000 m reached up to 9 $\mu g \cdot m^{-3}$. Only on 5 August, the ABL was well mixed, i.e. $PM_{2.5}$ concentrations were nearly constant with height. On August 6 and 7, 20 h mean $PM_{2.5}$ concentrations were highest at 500 m due to light rain showers from clouds with their highest ceilings at 250 m. This behavior did not occur for PM_{10}. Analysis suggested that some $PM_{2.5}$ swelled and converted to PM_{10}.

Overall, the 20 h mean $PM_{2.5}$ distributions derived from sampling followed those of PM_{10} throughout the episode at 200 m height (Figure 11, Figure 12). Throughout the episode, the greatest differences between the 20 h means from sampling and the "grand truth" occurred at the 200 m level. Here the plume had only marginal dispersion in the horizontal direction. With increasing height and time, the plume had expanded yielding more uniform $PM_{2.5}$ distributions as compared to the same area at 200 m height. As a result, differences between the distributions from sampling and the "grand truth" were smaller at the higher levels than at 200 m. However, when the plume migrated with time, notable differences occurred along the edges of the plume.

Generally, 20 h means of $PM_{2.5}$ from sampling overestimated the 20 h means of the "grand truth" (Figure 11). The default and second flight patterns typically produced similar 20 h mean $PM_{2.5}$ distributions with respect to the locations of highest concentrations. Typically, the second flight pattern suggested (up to 130 mg·m^{-3}) higher 20 h means of $PM_{2.5}$ than the default pattern. However, the minimum location differed for these flight patterns. The default pattern underestimated the 20 h mean $PM_{2.5}$ concentrations of the "grand truth" on average by 2 µg·m^{-3}. Analysis suggested that the substantial differences in 20 h mean $PM_{2.5}$ distribution were a product of timing of sampling and vertical mixing rather than horizontal transport.

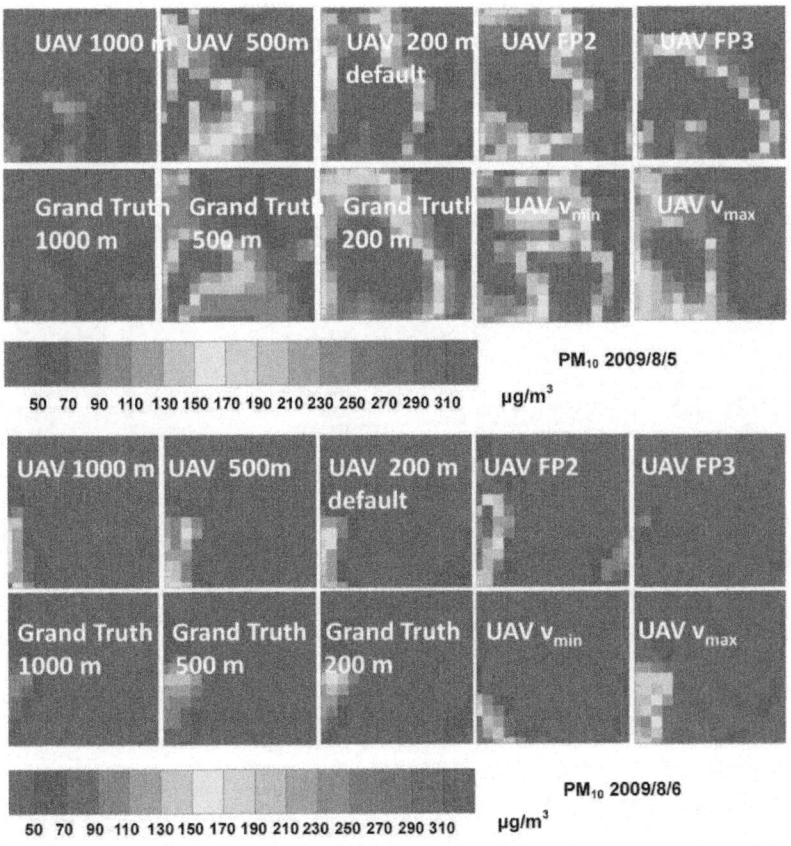

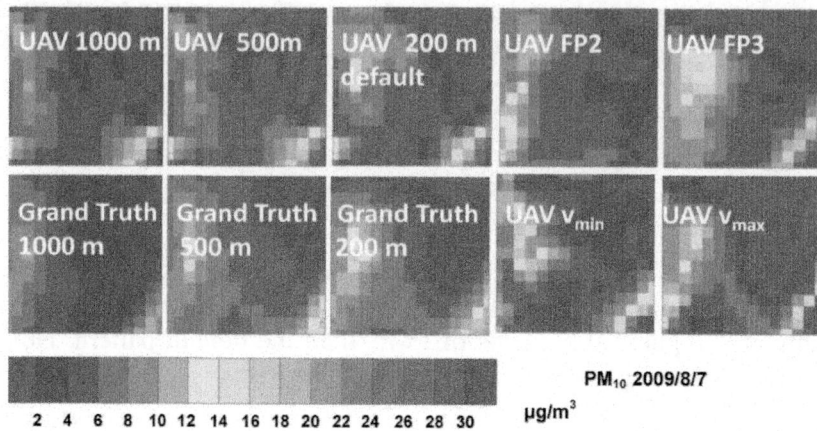

Figure 11. Distributions of mean PM_{10} concentrations as obtained (top left to bottom right) from virtual sampling at 1000 m, 500 m, 200 m (default), flight pattern 2 (FP2) at 200 m, flight pattern 3 (FP3) at 200 m, from the "grand truth" at 1000 m, 500 m, and 200 m, from sampling at stall speed (v_{min}) at 200 m, and maximum speed (v_{max}) at 200 m for prefrontal conditions on August 5 (first two rows of panel), conditions during the frontal passage on August 6 (middle two rows of panel) and after the front passed on August 7 (last two rows of panel). Panels show 20 h means for all sampling designs except for those using stall and maximum speed. Legend for August 7 differs from those for August 5 and 6.

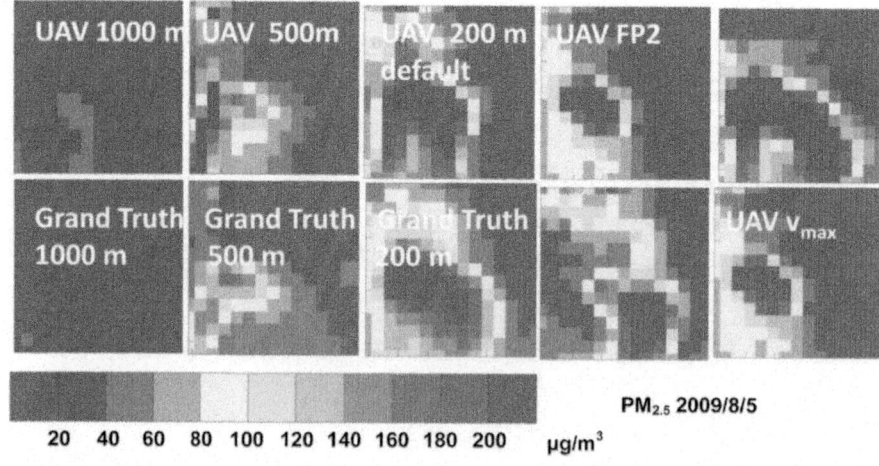

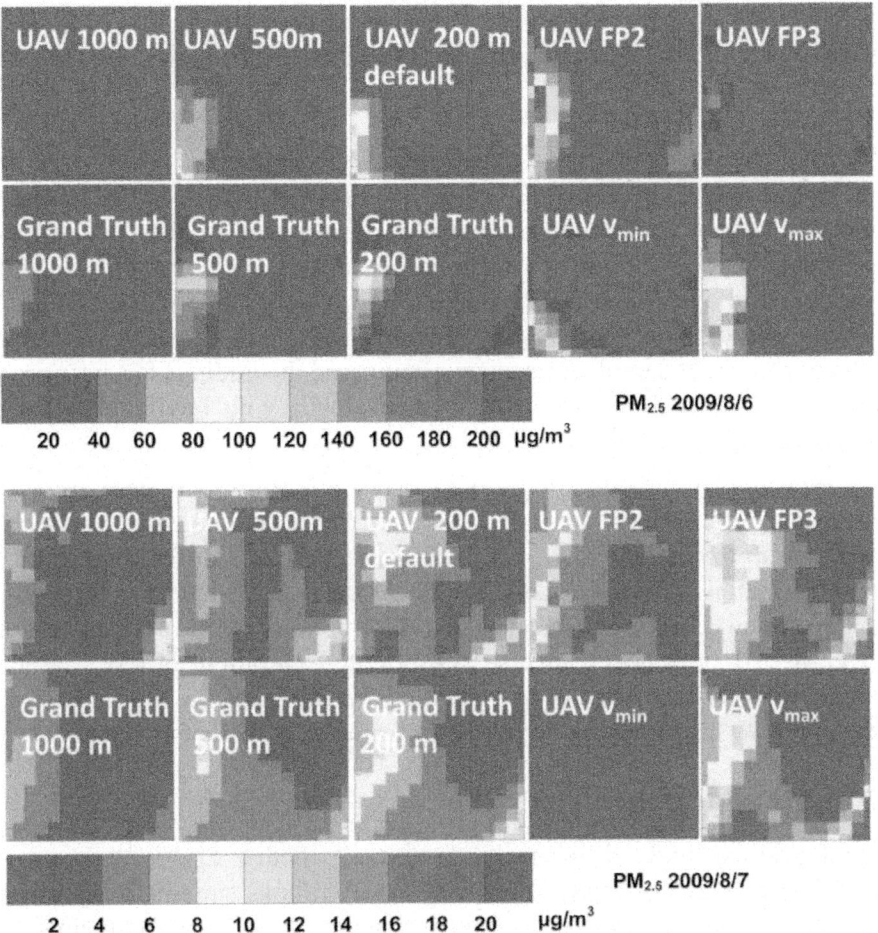

Figure 12. Distributions of mean $PM_{2.5}$ concentrations as obtained (top left to bottom right) from virtual sampling at 1000 m, 500 m, 200 m (default), flight pattern 2 (FP2) at 200 m, flight pattern 3 (FP3) at 200 m, from the "grand truth" at 1000 m, 500 m, and 200 m, from sampling at stall speed (v_{min}) at 200 m, and maximum speed (v_{max}) at 200 m for prefrontal conditions on August 5 (first two rows of panel), conditions during the frontal passage on August 6 (middle two rows of panel) and after the front passed on August 7 (last two rows of panel). Panels show 20 h means for all sampling designs except for those using stall and maximum speed. Legend for August 7 differs from those for August 5 and 6.

The largest differences among the 20 h mean $PM_{2.5}$ concentrations from sampling among each other and the "grand truth" occurred for stall speed (Figure 11). At cruising or maximum speed, the differences between the 20 h means from sampling showed more spatial variability than at stall speed.

Independent of flight speed, the 20 h mean $PM_{2.5}$ concentrations generally underestimated those of the "grand truth." Absolute mean fractional and normalized mean biases ranged from 2% to 66%. At maximum (minimum) speed, the overall mean FB and NMB were −2% (−66%) and −21% (−48%), respectively.

PM_{10} also encompasses $PM_{2.5}$. Comparison of PM_{10} and $PM_{2.5}$ concentrations showed that PM with diameters between 2.5 mm and 10 mm occurred at all three flight levels at all times. Due to the size dependency of settling velocities and of thermodynamic behavior under high relative humidity conditions, PM can experience stratification [66]. To understand the consequences of this behavior for UAV sampling results we compared the findings for these aerosol size classes. Stratification occurred when clouds with low ceilings formed below 250 m.

Typically, for both $PM_{2.5}$ and PM_{10} concentrations decreased with increasing height except for four cases when PM accumulated at the top of the ABL due to inversions. PM_{10} and $PM_{2.5}$ concentrations differed by about 0.1 $\mu g \cdot m^{-3}$ which is about the uncertainty of current state-of-the-art measurements on the ground. This finding means that (a) the majority of the particulate matter was $PM_{2.5}$ which is health adverse [69] ; (b) concentrations of $PM_{2.5}$ and PM_{10} were positively correlated independent of the choice of the flight pattern; (c) the same was true when the virtual UAV travelled at different speeds.

CONCLUSIONS

Our feasibility study theoretically examined whether UAVs could provide spatial distributions of mean pollutant concentrations suitable for public air-quality advisory. We used an episode during the 2009 Crazy Mountain fires in Interior Alaska as a test case.

Evaluated WRF/Chem data served to represent the conditions in the ABL and, hence, as "grand truth" in this study. A virtual Scan Eagle travelling at different heights, speeds and different patterns collected data from the WRF/Chem results along its flight path. We assumed optimum conditions with respect to flight duration, i.e. zero-weight payload and full fuel tank. Under such conditions, the Scan Eagle can fly about 20 h. The mean distributions derived from the sampled data were compared to the mean distributions according to the "grand truth." We examined a polluted situation with a fully developed wildfire-smoke plume, the removal of pollutants by a cold front passage, and the re-development of the smoke plumes. All quantities showed strong sensitivity to the flight patterns and heights on the day of the cold front passage.

Comparison of 20 h mean distributions obtained from sampling at different altitudes revealed the following: For air and dewpoint temperatures, differences were related to the environmental temperature lapse rate and the dewpoint temperature lapse rate, respectively. Concentrations of gases were nearly uniform with height under conditions of strong vertical mixing within the ABL.

In general, the virtual UAV captured the concentrations' reset to clean air background values after the cold front had passed the sampling domain, and the re-development of the plume thereafter. However, on the day of the frontal passage, 20 h mean distributions from sampling at different speeds and/or with different patterns led to different results and greater discrepancies from the 20 h means of the "grand truth" than found on the days prior to and after the frontal passage.

In the case of CO, the 20 h spatio-temporal variability obtained from sampling agreed with the "grand truth" within a factor of two at 1000 m, i.e. UAV sampling can provide good 20 h mean distributions of CO at 1000 m for 60 km ´ 60 km and retrieve information on smoke-plume propagation. Based on the results for CO, one may conclude, that in general, some of the 20 h mean pollutant concentrations obtained by the different flight patterns were due to changes in wind direction.

For primary pollutants involved in photochemical reaction chains (SO_2, NO) it is necessary to derive separate mean distributions for daytime and nighttime. The diurnal cycle of their concentrations led to overestimation (underestimation) of 20 h means in areas of high (low) concentrations as compared to the "grand truth." The unequal amount of daylight and dark hours and the time differences when the virtual UAV scanned a location caused that spatio-temporal biases did not cancel out.

Given the relative short darkness at high latitudes in late summer, the collection of enough nighttime data would require more than one UAV for coverage of the same area. Each UAV would have to fly a different sampling pattern for 20 h. The choice of sampling patterns must ensure that during darkness, the different UAVs would collect data in different areas of the sampling domain. The shorter the darkness, the more UAVs would be required which might cause logistic and personnel difficulties. The collected data would have to be sorted to create separate daylight and nighttime mean distributions.

The 20 h mean distributions of gases involved in photochemistry differed among flight patterns except for O_3. However, the 20 h mean O_3 distribution obtained from sampling depended on the speed of the virtual UAV.

The lowest possible safe flight height and cruising speed would provide information on how the underlying landscape modulates the smoke plume.

Sampling around the top of the ABL would provide information on the plume's dispersion and would be helpful for aviation advisory for small aircrafts when satellite imagery cannot provide this information due to a closed cloud cover in the mid- and upper troposphere. When deciding on the flight pattern, it is critical to consider wind speed, direction and precipitation, and their forecasted spatio- temporal evolutions during the planned flight duration. The sampling pattern must be designed to capture the conditions of interest (e.g. severity of pollution, washout due to frontal passage).

Sampling strategies for meteorological and chemical quantities might differ. Thus, air-quality forecasts and the virtual sampling technique introduced here may be an asset in effective, optimized flight planning, and collecting the data needed to answer the research question(s) at hand.

Our theoretical study assumed zero payload, i.e. a full tank. The heavier the payload the less fuel can be added, which reduces flight duration. While flight duration may be of low relevance for research questions related to short-term processes, deployment of UAVs for use in air-quality advisories requires a long flight duration to cover a large area in the downwind of the wildfire and for calculation of multi-hour means depending on the sampled species and its NAAQS averaging requirements. For some of the examined quantities, instruments light and small enough to fit in the UAV and that can sample at high enough frequency still have to be developed.

ACKNOWLEDGEMENTS

The authors wish to thank Uma S. Bhatt, R.L. Collins, M.C. Hatfield, D. Thorsen and the anonymous reviewers for fruitful discussion and helpful comments. The Research Support Center located at the Geophysical Institute of the University of Alaska Fairbanks provided CPU time and data storage. The National Aeronautics and Space Administration provided funding (Grant NASA-NNX11AQ27A).

REFERENCES

1.   Elston, J., Argrow, B., Stachura, M., et al. (2014) Overview of Small Fixed-Wing Unmanned Aircraft for Meteorological Sampling. Journal of Atmospheric and Oceanic Technology, 32, 97-115. http://dx.doi.org/10.1175/JTECH-D-13-00236.1

2.   US Census Bureau (2015) National Population Projections. https://www.census.gov/population/projections/data/national/2014.html

3.   PaiMazumder, D. and Molders, N. (2009) Theoretical Assessment of Uncertainty in Regional Averages Due to Network Density and Design.

Journal of Applied Meteorology and Climatology, 48, 1643-1666. http://dx.doi.org/10.1175/2009JAMC2022.1

4.   Shulski, M. and Wendler, G. (2007) The Climate of Alaska. University of Alaska Press, Fairbanks, 216 p.

5.   Bienek, P. (2007) Climate and Predictability of Alaska Wildfires. Master's Thesis, Department of Atmospheric Sciences, University of Alaska Fairbanks, Fairbanks, 95 p.

6.   Molders, N. and Kramm, G. (2007) Influence of Wildfire Induced Land-Cover Changes on Clouds and Precipitation in Interior Alaska—A Case Study. Atmospheric Research, 84, 142-168. http://dx.doi.org/10.1016/j.atmosres.2006.06.004

7.   Molders, N. (2008) Suitability of the Weather Research and Forecasting (WRF) Model to Predict the June 2005 Fire Weather for Interior Alaska. Weather and Forecasting, 23, 953-973. http://dx.doi.org/10.1175/2008WAF2007062.1

8.   Butwin, M.K. (2015) Theoretical Investigations on Strategies for Sampling Meteorological and Chemical Field Quantities in Smoke Plumes Using UAVs. Master's Thesis, Department of Atmospheric Sciences, University of Alaska Fairbanks, Fairbanks, 176 p.

9.   Mayer, S., Sandvik, A., Jonassen, M.O. and Reuder, J. (2010) Atmospheric Profiling with the UAS SUMO: A New Perspective for the Evaluation of Fine-Scale Atmospheric Models. Meteorology and Atmospheric Physics, 63, 15-26.

10.   Skamarock, W.C., Klemp, J.B., Dudhia, J., et al. (2008) A Description of the Advanced Research WRF Version 3. NCAR/TN, 125 p.

11.   Elston, J.S., Roadman, J., Stachura, M., et al. (2011) The Tempest Unmanned Aircraft System for in Situ Observations of Tornadic Supercells: Design and VORTEX2 Flight Results. Journal of Field Robotics, 28, 461-483. http://dx.doi.org/10.1002/rob.20394

12.   Patterson, M.C.L., Mulligan, A., Douglas, J., et al. (2005) Volcano Surveillance by ACR Silver Fox. American Institute of Aeronautics and Astronautics, 1, 26-29. http://dx.doi.org/10.2514/6.2005-6954

13.   McGonigle, A.J.S., Aiuppa, A., Giudice, G., et al. (2008) Unmanned Aerial Vehicle Measurements of Volcanic Carbon Dioxide Fluxes. Geophysical Research Letters, 35, Article ID: L06303. http://dx.doi.org/10.1029/2007gl032508

14.   EPA (2011) National Ambient Air Quality Standards (NAAQS). http://www.epa.gov/air/criteria.html

15. Grell, G.A., Peckham, S.E., Schmitz, R., et al. (2005) Fully Coupled "Online" Chemistry within the WRF Model. Atmospheric Environment, 39, 6957-6975.http://dx.doi.org/10.1016/j.atmosenv.2005.04.027

16. Peckham, S.E., Fast, J., Schmitz, R., et al. (2011) WRF/Chem Version 3.3 User's Guide. 96 p.

17. Mlawer, E.J., Taubman, S.J., Brown, P.D., Iacono, M.J. and Clough, S.A. (1997) Radiative Transfer for Inhomogeneous Atmospheres: RRTM, a Validated Correlated-K Model for the Longwave. Journal of Geophysical Research, 102D, 16663-16682.http://dx.doi.org/10.1029/97JD00237

18. Chou, M.-D. and Suarez, M.J. (1994) An Efficient Thermal Infrared Radiation Parameterization for Use in General Circulation Models. Report, Goddard Space Flight Center, Greenbelt, 85 p.

19. Grell, G.A. and Dévényi, D. (2002) A Generalized Approach to Parameterizing Convection. Geophysical Research Letters, 29, 1693.

20. Lin, Y.-L., Rarley, R.D. and Orville, H.D. (1983) Bulk Parameterization of the Snow Field in a Cloud Model. Journal of Applied Meteorology, 22, 1065-1092. http://dx.doi.org/10.1175/1520-0450(1983)022<1065:BPOTSF>2.0.CO;2

21. Barnard, J., Fast, J., Paredes-Miranda, G., Arnott, W. and Laskin, A. (2010) Technical Note: Evaluation of the WRF-Chem "Aerosol Chemical to Aerosol Optical Properties" Module Using Data from the MILAGRO Campaign. Atmospheric Chemistry and Physics, 10, 7325-7340. http://dx.doi.org/10.5194/acp-10-7325-2010

22. Grell, G.A., Freitas, S.R., Stuefer, M. and Fast, J.D. (2011) Inclusion of Biomass Burning in WRF-Chem: Impact on Wildfires on Weather Forecasts. Atmospheric Chemistry Physics, 11, 5289-5303. http://dx.doi.org/10.5194/acp-11-5289-2011

23. Mellor, G.L. and Yamada, T. (1982) Development of a Turbulence Closure Model for Geophysical Fluid Problems. Review of Geophysics—Space Physics, 20, 851-875. http://dx.doi.org/10.1029/RG020i004p00851

24. Janjic, Z.I. (1994) The Step-Mountain Eta Coordinate Model: Further Developments of the Convection, Viscous Sublayer and Turbulence Closure Schemes. Monthly Weather Review, 122, 927-945. http://dx.doi.org/10.1175/1520-0493(1994)122<0927:TSMECM>2.0.CO;2

25. Smirnova, T.G., Brown, J.M., Benjamin, S.G. and Kim, D. (2000) Parameterization of Cold Season Processes in the Maps Land-Surface Scheme. Journal Geophysical Research, 105D, 4077-4086. http://dx.doi.

org/10.1029/1999JD901047

26. Stockwell, W.R., Middleton, P., Chang, J.S. and Tang, X. (1990) The Second-Generation Regional Acid Deposition Model Chemical Mechanism for Regional Air Quality Modeling. Journal Geophysical Research, 95, 16343-16367.http://dx.doi.org/10.1029/JD095iD10p16343

27. Madronich, S. (1987) Photodissociation in the Atmosphere, 1, Actinic Flux and the Effects of Ground Reflections and Clouds. Journal Geophysical Research, 92, 9740-9752. http://dx.doi.org/10.1029/JD092iD08p09740

28. Wesely, M.L. (1989) Parameterization of Surface Resistances to Gaseous Dry Deposition in Regional-Scale Numerical Models. Atmospheric Environment, 23, 1293-1304. http://dx.doi.org/10.1016/0004-6981(89)90153-4

29. Molders, N., Tran, H.N.Q., Quinn, P., et al. (2011) Assessment of WRF/Chem to Capture Sub-Arctic Boundary Layer Characteristics during Low Solar Irradiation Using Radiosonde, Sodar, and Station Data. Atmospheric Pollution Research, 2, 283-299.http://dx.doi.org/10.5094/APR.2011.035

30. Ackermann, I.J., Hass, H., Memmesheimer, M., Ziegenbein, C. and Ebel, A. (1995) The Parametrization of the Sulfate-Nitrate-Ammonia Aerosol System in the Long-Range Transport Model EURAD. Meteorology and Atmospheric Physics, 57, 101-114.http://dx.doi.org/10.1007/BF01044156

31. Ackermann, I.J., Hass, H., Memmesheimer, M., et al. (1998) Modal Aerosol Dynamics Model for Europe: Development and First Applications. Atmospheric Environment, 32, 2981-2299. http://dx.doi.org/10.1016/S1352-2310(98)00006-5

32. Kramm, G., Beheng, K.-D. and Müller, H. (1992) Vertical Transport of Polydispersed Aerosol Particles in the Atmospheric Surface Layer. In: Schwartz, S.E. and Slinn, W.G.N., Eds., Precipitation Scavenging and Atmosphere-Surface Exchange Processes, Vol. 2, Hemisphere Publishing Company, Washington/Philadelphia/London, 1125-1141.

33. Schell, B., Ackermann, I.J., Hass, H., Binkowski, F.S. and Ebel, A. (2001) Modeling the Formation of Secondary Organic Aerosol within a Comprehensive Air Quality Model System. Journal of Geophysical Research, 106, 28275-28293.http://dx.doi.org/10.1029/2001JD000384

34. Freitas, S.R., Longo, K.M., Alonso, M.F., et al. (2011) Prep-Chem-SRC—1.0: A Preprocessor of Trace Gas and Aerosol Emission Fields for Regional and Global Atmospheric Chemistry Models. Geoscientific

Model Development, 4, 419-433.http://dx.doi.org/10.5194/gmd-4-419-2011

35. NASA EOSDIS (2012) Near Real-Time Data. http://lance-modis.eosdis.nasa.gov/imagery/subsets/?area=global

36. Olson, J., Watts, S. and Allison, L.J. (2000) Major World Ecosystem Complexes Ranked by Carbon in Live Vegetation: A Database (Revised November 2000). Carbon Dioxide Information Analysis Center, Oak Ridge National Laboratory, Oak Ridge.

37. van Aardenne, J.A., Dentener, F., Olivier, J.G.J. and Peters, J.A.H.W. (2005) The EDGAR 3.2 Fast Track 2000 Dataset. (32FT2000)

38. Molders, N. (2010) Alaska Emission Model (AkEM)—Version 1.01 Description. Internal Report, Fairbanks, 16 p.

39. Guenther, A., Hewitt, C., Erickson, D., et al. (1994) A Global Model of Natural Volatile Organic Compound Emissions. Journal Geophysical Research, 100D, 8873-8892.

40. Simpson, D., Guenther, A., Hewitt, C.N. and Steinbrecher, R. (1995) Biogenic Emissions in Europe 1. Estimates and Uncertainties. Journal Geophysical Research, 100D, 22875-22890. http://dx.doi.org/10.1029/95JD02368

41. Department of Commerce (2000) NCEP FNL Operational Model Global Tropospheric Analyses, Continuing from July 1999. Research Data Archive at the National Center for Atmospheric Research, Computational and Information Systems Laboratory.http://rda.ucar.edu/datasets/ds083.2/

42. Sassen, K. (1991) The Polarization Lidar Technique for Cloud Research: A Review and Current Assessment. Bulletin of the American Meteorologlogical Society, 72, 1848-1866. http://dx.doi.org/10.1175/1520-0477(1991)072<1848:TPLTFC>2.0.CO;2

43. Mishchenko, M.I. and Sassen, K. (1998) Depolarization of Lidar Returns by Small Ice Crystals: An Application to Contrails. Geophysical Research Letters, 25, 309-312. http://dx.doi.org/10.1029/97GL03764

44. Sassen, K. (2000) Lidar Backscatter Depolarization Technique for Cloud and Aerosol Research. In: Mishchenko, M.I., Hovenier, J.W. and Travis, L.D., Eds., Light Scattering by Nonspherical Particles, Academic Press, San Diego, 393-416.http://dx.doi.org/10.1016/B978-012498660-2/50041-0

45. Sassen, K. (2005) Dusty Ice Clouds over Alaska. Nature, 434, 456.http://dx.doi.org/10.1038/434456a

46. Madden, J.M. (2014) Using WRF/Chem, In-Situ Observations, and CALIPSO Data to Simulate Smoke Plume Signatures on High-Latitude Pixels. Master's Thesis, Department of Atmospheric Sciences, University of Alaska Fairbanks, Fairbanks, 106 p.

47. Madden, J.M., Molders, N. and Sassen, K. (2015) Assessment of WRF/Chem Simulated Vertical Distributions of Particulate Matter from the 2009 Minto Flats South Wildfire in Interior Alaska by CALIPSO Total Backscatter and Depolarization Measurements. Open Journal of Air Pollution, 4, 119-138. http://dx.doi.org/10.4236/ojap.2015.43012

48. Sassen, K. (2008) Boreal Tree Pollen Sensed by Polarization Lidar: Depolarizing Biogenic Chaff. Geophysical Research Letters, 35, Article ID: L18810.http://dx.doi.org/10.1029/2008gl035085

49. Martins, J.V., Hobbs, P.V., Weiss, R.E. and Artaxo, P. (1998) Sphericity and Morphology of Smoke Particles from Biomass Burning in Brazil. Journal of Geophysical Research—Atmosphere, 103, 32051-32057. http://dx.doi.org/10.1029/98JD01153

50. Murayama, T., Müller, D., Wada, K., et al. (2004) Characterization of Asian Dust and Siberian Smoke with Multi-Wavelength Raman Lidar over Tokyo, Japan in Spring 2003. Geophysical Research Letters, 31, Article ID: L23103.http://dx.doi.org/10.1029/2004GL021105

51. Sassen, K. (2008) Identifying Atmospheric Aerosols with Polarization Lidar. In: Kim, Y.J. and Platt, U., Eds., Advanced Environmental Modeling, Springer Science + Business Media Inc., New York, 136-142. http://dx.doi.org/10.1007/978-1-4020-6364-0_10

52. Wandinger, U., Müller, D., Bockmann, C., et al. (2002) Optical and Microphysical Characterization of Biomass-Burning and Industrial-Pollution Aerosols from Multiwavelength Lidar and Aircraft Measurements. Journal of Geophysical Research, 107, 8125-8145. http://dx.doi.org/10.1029/2000JD000202

53. Lee, C.H., Kim, J.H., Park, C.B., et al. (2004) Continuous Measurements of Smoke of Russian Forest Fire by 532/1064 nm Mie Scattering Lidar at Suwon, Korea. In: Proceedings of the 22nd International Laser Radar Conference, European Space Agency, Paris, 535-538.

54. Hu, Y.M., Vaughan, M., Liu, Z., et al. (2007) The Depolarization—Attenuated Backscatter Relation: CALIPSO Lidar Measurements vs. Theory. Optical Express, 15, 5327-5332. http://dx.doi.org/10.1364/OE.15.005327

55. Chang, J.C. and Hanna, S.R. (2004) Air Quality Model Performance Evaluation. Meteorology and Atmospheric Physics, 87, 167-196. http://dx.doi.org/10.1007/s00703-003-0070-7

56. EPA (2007) Guidance on the Use of Models and Other Analyses for Demonstrating Attainment of Air Quality Goals for Ozone, PM2.5, and Regional Haze. 262 p. http://www3.epa.gov/scram001/guidance/guide/final-03-pm-rh-guidance.pdf

57. Appel, K., Roselle, S., Gilliam, R. and Pleim, J. (2010) Sensitivity of the Community Multiscale Air Quality (CMAQ) Model v4.7 Results for the Eastern United States to MM5 and WRF Meteorological Drivers. Geoscience Model Development, 3, 169-188.http://dx.doi.org/10.5194/gmd-3-169-2010

58. Hines, K.M. and Bromwich, D.H. (2008) Development and Testing of Polar Weather Research and Forecasting (WRF) Model. Part I: Greenland Ice Sheet Meteorology. Monthly Weather Review, 136, 1971-1989. http://dx.doi.org/10.1175/2007MWR2112.1

59. Hines, K.M., Bromwich, D.H., Bai, L.-S., Barlage, M. and Slater, A.G. (2011) Development and Testing of Polar WRF. Part III. Arctic Land. Journal of Climate, 24, 26-48. http://dx.doi.org/10.1175/2010JCLI3460.1

60. Avissar, R. and Pielke, R.A. (1989) A Parameterization of Heterogeneous Land Surface for Atmospheric Numerical Models and Its Impact on Regional Meteorology. Monthly Weather Review, 117, 2113-2136. http://dx.doi.org/10.1175/1520-0493(1989)117<2113:APOHLS>2.0.CO;2

61. Molders, N. and Raabe, A. (1996) Numerical Investigations on the Influence of Subgrid-Scale Surface Heterogeneity on Evapotranspiration and Cloud Processes. Journal of Applied Meteorology, 35, 782-795. http://dx.doi.org/10.1175/1520-0450(1996)035<0782:NIOTIO>2.0.CO;2

62. Molders, N., Tran, H.N.Q., Cahill, C.F., Leelasakultum, K. and Tran, T.T. (2012) Assessment of WRF/Chem PM2.5 Forecasts Using Mobile and Fixed Location Data from the Fairbanks, Alaska Winter 2008/09 Field Campaign. Air Pollution Research, 3, 180-191.http://dx.doi.org/10.5094/apr.2012.018

63. Pirhalla, M.A., Gende, S. and Molders, N. (2014) Fate of Particulate Matter from Cruise-Ship Emissions in Glacier Bay during the 2008 Tourist Season. Journal of Environmental Protection, 4, 1235-1254. http://dx.doi.org/10.4236/jep.2014.512118

64. von Storch, H. and Zwiers, F.W. (1999) Statistical Analysis in Climate

Research. Cambridge University Press, Cambridge, 484 p.

65. Molders, N. and Kramm, G. (2014) Lectures in Meteorology. Heidelberg, Springer, 591 p.

66. Seinfeld, J.H. and Pandis, S.N. (1997) Atmospheric Chemistry and Physics, from Air Pollution to Climate Change. John Wiley & Sons, New York, 1326 p.

67. Hanna, S. and Chang, J. (2012) Acceptance Criteria for Urban Dispersion Model Evaluation. Meteorology and Atmospheric Physics, 116, 133-146. http://dx.doi.org/10.1007/s00703-011-0177-1

68. Freitas, S.R., Longo, K.M., Silva Dias, M.A.F. and Artaxo, P. (1996) Numerical Modeling of Air Mass Trajectories from the Biomass Burning Areas of the Amazon Basin. Annals of the Brazilian Academy of Sciences, 68, 193-206.

69. Pope, I.C.A., Dockery, D.W. and Schwartz, J. (1995) Review of Epidemiological Evidence of Health Effects of Particulate Air Pollution. Inhalation Toxicology, 7, 1-18.

Chapter 7

EFFECTIVENESS OF AN INDOOR AIR POLLUTION (IAP) INTERVENTION ON REDUCING IAP AND IMPROVING WOMEN'S HEALTH STATUS IN RURAL AREAS OF GANSU PROVINCE, CHINA

Yibin Cheng[1], Jiaqi Kang[1], Fan Liu[1], Bryan A. Bassig[2], Brian Leaderer[2], Gongli He[1], Theodore R. Holford[2], Ning Tang[1], Jian Wang[3], Jian He[4], Yanchang Liu[1], Yingchun Liu[1], Jiang Liu[1], Xun Chen[1], Heng Gu[1], Xiao Ma[5], Tongzhang Zheng[2], Yinlong Jin[1]

[1]Institute for Environmental Health and Related Product Safety, Chinese Center for Disease Control and Prevention, Beijing, China

[2]Yale School of Public Health, New Haven, USA

[3]Chinese Center for Disease Control and Prevention, Beijing, China

[4]Gansu Provincial Center for Disease Control and Prevention, Lanzhou, China

[5]West China School of Public Health, Sichuan University, Chengdu, China

ABSTRACT

Given the deleterious health effects associated with indoor air pollution (IAP), this study was conducted to evaluate an IAP intervention in rural areas in Gansu, one of the poorest provinces of China. We selected 371 rural households to take part in intervention measures including stove improvement and health education. Eight of 371 households were selected to conduct IAP sampling. Four hundred and thirteen women in these households completed a questionnaire and 49 women took part in lung function tests. After the intervention, PM_4 levels reduced from 455 $\mu g/m^3$ to 200 $\mu g/m^3$ and CO reduced from 3.40 ppm to 2.90 ppm in indoor air. The percentage of predicted value of FEV1 and FVC improved to some degree after the intervention, but all the parameters of lung function assessment did not show a significant change. Prevalence rates of several symptoms associated with IAP significantly declined in the study population, compared with baseline levels. Intervention measures combining stove improvement with health education were effective in reducing IAP levels. Women's health status, including eye and respiratory symptoms, also showed improvement. However, the effect on lung function was not apparent and warranted additional follow-up. Similarly, evaluation of the long term effects of the IAP intervention will require future studies.

INTRODUCTION

About 41% of the world's population use solid fuel (such as coal and biomass) for domestic cooking and heating, and a large proportion of this exposed population lives in less developed countries [1]. Globally, reliance on solid fuels has emerged as one of the ten most important threats to public health [2]. In 2010, household air pollution from solid fuels was among the top 3 risk factors contributing to the global burden of disease and was responsible for nearly 3.5 million deaths across the world [3]. These deaths occur predominantly in women and children, as women are normally responsible for food preparation and cooking, and infants and young children are usually with their mothers near the cooking area and therefore are more likely to be exposed. In addition to deaths, various health issues, such as respiratory diseases, lung function reduction, and eye infections, have been associated with IAP exposure as recently reviewed or reported by others [4] - [6]. For example, exposure to high levels of SO_2 in the kitchen has been associated with a significantly higher prevalence of COPD among nonsmoking women [7]. Furthermore, women living in households that used biomass had a significantly higher prevalence of asthma than those in households using cleaner fuels, and the risk for asthma appeared to be higher for women than for men [8].

More than half of the Chinese population lives in rural areas where more than 80% of the households use solid fuels for cooking [9]. WHO estimates of the disease burden associated with this use in China had suggested that solid fuel use was responsible for about 342,450 COPD deaths and 17,720 lung cancer deaths for those aged ≥ 30, and was also responsible for 3,204,900 total DALYs in China in 2002, which was equivalent to 1.6% of the national disease burden during that year [2]. A further 28% of global deaths caused by indoor smoke from solid fuels occur in China [10].

In 2002, the World Bank, in conjunction with the Institute of Environmental Health and Related Product Safety of the Chinese Center for Disease Control and Prevention (IEHS, China CDC), initiated an IAP intervention product called the Sustainable and Efficient Energy Use to Alleviate Indoor Air Pollution in Poor Rural Areas of China. The purpose of the project was to test the viability of both technological and behavioral interventions to mitigate IAP and improve human health, and the project was carried out in rural areas of 4 Chinese remote provinces including Gansu, Guizhou, Shaanxi, and Inner Mongolia. The IAP intervention involved stove improvement in each selected area and health education (behavior intervention) of the local residents with the goal of changing their traditional cooking methods, heating practices, and other lifestyles that might contribute to IAP [11] [12].

Some previous epidemiologic evidence has indicated the beneficial effects of stove improvement as far as associated reduced risks for lung cancer, pneumonia, and COPD in other rural areas in China following implementation, which suggests that such an intervention may be viable [13] - [15]. Here, we report the results assessing the effectiveness of the IAP intervention on reducing IAP and improving the health status in women in Hui County of Gansu Province. We report the effects of the intervention on multiple indices related to IAP mitigation, including changes in various symptoms of eye and respiratory diseases, measures of lung function, and assessment of PM_4 and CO levels in a representative sample of the households both before and after the intervention period.

MATERIALS AND METHODS

Study Population

Gansu, located in northwest China, is one of China's least developed provinces. The annual income per capita for rural residents in the province was estimated to be around US $193 in 2002 according to the National Bureau of Statistics of China. Wood and crop residues are the main energy sources for domestic cooking and heating in this area. The regular heating season in the region is between November and April of the following year. Hui County was selected for the stove intervention study because households in this county commonly use traditional stoves with biomass for heating and cooking, which causes direct release of smoke into the indoor living environment. While some households had chimneys, the chimneys were installed too low (for example, lower than eave) and consequently the smoke can easily reenter into rooms and result in IAP. Moreover, very few families in the area installed ventilators in their kitchen.

A total of 413 women from 371 households were included in the stove intervention study. Households that met the following criteria were recruited for this study: 1) used biomass for heating and cooking before the intervention with traditional stoves, e.g. open fire, unvented stoves, or stoves without a smoke door; 2) included woman aged 18 and over and children under age 15; 3) had residents who lived in Hui County continuously for more than one year at the time that the stove intervention began.

Stove Improvement

Stove types used in homes in this region before the intervention consisted of either open pits (Figure 1(a)) or hand stoves (Figure 1(b)), or traditional stoves with either no chimney in the home (Figure 1(c)) or a chimney that was installed too low to effectively mitigate IAP (Figure 1(d)). The improved stoves with a chimney that were used for the intervention (Figure 2(a) and Figure 2(b)) were specifically designed for the local residents and were pilot tested for effectiveness before the stove intervention, as described elsewhere [16]. The local technicians were trained and certified by the Chinese NIEHS and were responsible for the installation of the stoves and for training each of the residents on how to use the stoves. Because some of the traditional stoves were used by the subjects for heating, the improved stove was connected to a heating bed (called Kang as shown inFigure 2(c)) and thus could be used for heating purposes as well. Results from the pilot tests showed that emission of PM_4 and CO reduced by 98% and 88%, respectively, in controlled conditions [16]. Between August and November in 2004, all 371 participating families changed to a certified stove with a chimney for cooking and heating. The old stoves inside these homes were completely demolished and replaced by the improved stoves. All homes in this region use agricultural residue and wood as their primary source of fuel for cooking and heating, and these fuel types did not change over the course of this study.

(a)

(b)

(c)

(d)

Figure 1. (a)-(d) Stoves used by the households in the study area before the intervention. Stoves used in this region before the intervention included either: open pits (a) or hand stoves (b), or traditional stoves with either no chimney in the home (c) or a chimney that was installed too low to effectively mitigate IAP (d).

(a)

(b)

(c)

Figure 2. Improved stoves used by the households in the study area after the intervention. The improved stoves with a chimney that were used for the intervention (a) and (b) were specifically designed for the local residents. The improved stove was connected to a heating bed (called Kang as shown in (c).

Health Education (Behavior Intervention)

To encourage residents to actively participate in the stove improvement activities and correctly use the newly installed stoves, and to change lifestyle habits that may contribute to IAP or increasing exposure to IAP, scientists from IEHS, China CDC, and the Huaxi School of Public Health, Sichuan University, trained and worked with the local health professionals, village physicians, local stove technicians, and school teachers to educate the local residents during 2003 and 2005. This component included providing knowledge on the sources and impacts of IAP and information on correctly using and maintaining the stoves. Pamphlets describing methods to prevent IAP were distributed during household visits, health education classes, and other related activities. Three specific factors were evaluated to measure the impact of the behavior intervention, including use of a fan in the kitchen, opening of windows during the winter, and reduction in use of portable open fires to warm the pot tea.

Assessment of Health and Intervention Effects

Following informed consent, information on respiratory symptoms and occurrence of eye irritation/infections were collected through in-person interviews conducted by 30 trained interviewers. In both pre-intervention baseline interviews (during April, 2003) and post-intervention interviews (during April, 2005), participating women were asked to report their respiratory and eye symptoms during the 3 months before the interviews. The post- intervention health assessments, including lung function testing, were conducted between 5 - 8 months after the improved stoves were installed during August-November 2004. Information on the type and amount of energy use, stove type, cooking habits, ventilation, heating, and other lifestyle and demographic data was also collected during the in-person interviews. A total of 49 women participated in both pre- and post-intervention lung function testing. These 49 women were similar to the overall study population with respect to demographic characteristics and cooking practices, as well as characteristics associated with the home, such as house area and presence of a short chimney. The pre-intervention spirometry was performed in April 2003 and post-intervention spirometry was conducted in April 2005 by technicians from IEHS, China CDC using a Multi-Functional Spirometer HI-801 (Chest M.I., INC, Tokyo, Japan). These technicians were trained before the field component of the study in accordance with the standardized steps of the ATS/ERS Task Force: Standardization of Lung Function Testing. The collected measurements on lung function parameters included vital capacity (VC), forced expiratory volume in 1 second (FEV1), forced vital capacity (FVC), and forced expiratory flow 25% - 75% (FEF 25% - 75%). All lung function tests were conducted by trained personnel in the morning during the study period. For each of the lung function parameters, three measurements were taken and the best result of the 3 measurements was recorded and reported here.

Assessment of Indoor Air Pollutants

Because of the cost of monitoring and the willingness of the households, eight households were selected for IAP monitoring. These 8 families were selected to represent the geographic location, stove type, and type of biomass used in the region. PM_4 and CO in the kitchen and bedroom were measured before (March 2003) and between 5-8 months after the stove intervention (March 2005). Indoor air samples were collected consecutively for 24 hours in each of the 8 households. Respirable particles were measured according to the National Institute for Occupational Safety and Health of the United States (NIOSH, USA) protocol 0600, designed to capture particles with a median aerodynamic diameter of 4 ím (PM4). Samples were collected using a 10-mm nylon cyclone equipped with a 37-mm diameter poly vinyl chloride (PVC) filter (pore size 5ím supplied by SKC Inc., USA) at a flow rate of 2.5 l/min. Air was drawn through the cyclone preselectors using battery-operated constant flow pumps (model PCXR8 supplied by SKC Inc., USA). All pumps were calibrated prior to and after each sampling day using a field minimeter, itself calibrated by a soap bubble meter in the laboratory. Pumps were also calibrated in the laboratory after each field exercise using the same minimeter. To maintain battery power throughout the sampling period, pumps were programmed to cover the 24-h interval through intermittent sampling (1 min out of every 4 - 6 min). One field blank was taken on each sampling day.

Gravimetric analyses were conducted at the laboratory of the National Institute for Environmental Health and Related Products Safety, China CDC using an analytic microbalance (1/100,000, Sartorius 2004 MP, Germany) calibrated against standards provided by the Bureau of National Technological Control. All filters (field blanks and samples) were conditioned for 24 hours before weighing. The weighing facility temperature was set at 20 - 25 degrees C and humidity at 50% ± 5%. Respirable dust concentrations were calculated by dividing the blank- corrected increase in filter mass by the total air volume sampled [17]. Carbon monoxide (CO) was measured using long term diffusion tubes (manufactured by GASTEC, USA), with detection ranges of 10 - 200 or 50 - 1000 ppm [17].

Statistical Analysis

The 24 hour personal exposure to PM_4 and CO in kitchen and bedroom air was compared for the pre- and post- IAP intervention periods using the Wilcoxon signed rank test. Changes in lifestyles before and after the intervention were compared using McNemar's test. Lung function was assessed by calculating means and standard deviations based on observed values and the percentage of predicted values of each parameter. Predicted values were the average observed values in the population for any person of similar age, sex, and body composition, calculated in terms of predict equation for the Chinese population [18]. The percentages of predicted values were the ratios of observed values divided by the predicted value so that age, sex, and body composition would be adjusted by comparing the relative values. In addition to the possible change of body composition, we used linear mixed models to adjust for other factors including cooking years up to the date of survey, use of an open fire, not using fans in the kitchen, location of the kitchen in the home, and whether or not windows were opened when cooking. For the total study population (413 women), the effectiveness of the IAP intervention on reducing the respiratory and eye irritation symptoms was evaluated by logistic regression, adjusting for age, cooking years up to the date of the survey, using open fire, location of the kitchen, and never opening windows when cooking. For 49 spirometry participants, McNemar's test was used for the analysis of symptoms.

RESULTS

Characteristics of Study Population

Characteristics of the study population based on data from the baseline survey are shown in Table 1. Over 80% of the subjects were between 18 - 39 years old and 99.8% of subjects were Han, the main ethnicity of China. Over 50% of the participants lived in households with 5 or more family members, and biomass was the main living fuel in all the selected households. 98.1% of women were responsible for cooking and 63.7% of subjects reported an average daily cooking time of 2 to 4 hours. None of the subjects were smokers, but 70% of them reported exposure to second-hand smoke. Very few women reported a change in their passive smoking exposure status after the intervention, compared to before because the interviews took place after such a short period following the stove improvement. The prevalence rates for all of the considered chronic diseases were low and were less than 5% in the study population.

Changes of Stove Features and Behaviors

The changes in stove features and behaviors relating to stove use associated with the intervention are shown in Table 2. Before the intervention, the percentages of stoves without chimneys, short chimneys (i.e., smoke exit lower than eaves), and stoves without smoke doors were 6.1%, 71.7% and 87.7%, respectively. These traditional designs of the stoves were considered primary contributors to IAP in households that used biomass. Following the intervention, the percentages of stoves having each of these characteristics reduced to 0% (Table 2).

Table 1. Characteristics of the study population (%).

Variables	n	Percentage %
Age (baseline)		
18 - 29	109	26.4
30 - 39	230	55.7
40 - 49	32	7.8
50 - 59	31	7.5
60~	11	2.7
Ethnicity		
Han	412	99.8
Other	1	0.2
Family members		
3	45	10.9
4	135	32.7
5	112	27.1
6	84	20.3
>6	36	8.7
Missing	1	0.2
Biomass use	413	100.0
Cooking	405	98.1
Everyday cooking time (hrs)		
<2	105	25.4
2~	263	63.7
4~	39	9.4
6~	6	1.5
Smoking	0	0
Passive smoking	289	70.0
History of diagnosed chronic diseases (%)		
Rhinitis	3	0.7

Faucitis	2	0.5
Tuberculosis	2	0.5
Asthma	1	0.2
Emphysema	0	0
Chronic bronchitis	9	2.2
Hypertension	12	2.9
Heart disease	14	4.1
Allergy	4	1.0

Table 2. Changes of stove features and behaviors before and after intervention (%).

Stove features and behaviors	Before (n = 413)	After (n = 413)	p-value[b]
Stoves without chimneys	6.1	0	–
Short chimney (exit of chimney lower than eave)[a]	71.7	0	–
Stoves without smoke doors	87.7	0	–
Portable open fire used for heating tea	35.6	33.7	0.01
Fans installed in kitchen	1.0	5.6	<0.01
Never open window in winter	16.0	11.1	0.048

[a]Twenty-five records were missing; [b]McNemar's Test.

Alteration of behaviors associated with IAP exposure comparing the pre and post intervention periods are additionally shown in Table 2. These behaviors were compared to assess the health education component of the intervention. Using portable open fires to warm the pot tea, a tradition kept for hundreds of years in this region, decreased from 35.6% before the intervention to 33.7% after the intervention (p = 0.01). During the 2-year intervention, more households chose to buy and equip fans in kitchens to exhaust cooking smoke, and this use increased from 1.0% to 5.6% (p < 0.01) among enrolled participants. Furthermore, the percentage of participants who never opened windows during winter reduced from 16.0% to 11.1% over the course of the intervention period (p = 0.048; Table 2).

Improvement of Indoor Air Quality

Households for IAP sampling were selected from the 371 households, considering house area, type of fuel used, stove features, and ventilation on the basis of a pilot study. Classifying by quantiles in accordance with questionnaire data of 413 women, the sample distribution of household areas was similar between the total population and the selected 8 households. Overall, the characteristics of the households were similar comparing the total study population and those households selected for sampling (Table 3), and in addition sociodemographic characteristics including income and education level were similar among women in households with and without IAP monitoring.

A comparison of PM_4 and CO levels before and after the intervention stratified by room type is shown in Table 4 and Table 5. The average concentration of PM_4 decreased by about 70%, from 774 to 223 $\mu g/m^3$, in the kitchen following implementation of the intervention, though this effect was marginally significant (p = 0.08) due to the small sample size. Conversely, there was no observed decrease in the average bedroom concentrations of PM_4 following the intervention (pre 135 $\mu g/m^3$vs. post 176 $\mu g/m^3$; p = 0.20), whereas the average PM_4 levels of the bedroom and kitchen combined were decreased by about 56% following the intervention (pre 455 $\mu g/m^3$ vs. post 200 $\mu g/m^3$; p = 0.46). The concentration of PM_4 in the bedrooms was lower than that in the kitchen both before and after stove improvement (Table 4).

The average concentrations of CO in both the kitchen and bedroom were reduced following the intervention (Table 5). Specifically, CO levels decreased in the kitchen from 3.81 ppm to 3.00 ppm, while concentrations reduced from 2.99 ppm to 2.80 ppm in the bedroom. For both rooms combined, CO levels changed slightly, declining from 3.40 ppm to 2.90 ppm, although none of these reductions of CO were statistically significant (Table 5).

Changes of Health Indicators of Women

Changes in lung function following the intervention are reported in Table 6. Spirometry of women showed that FEV1 increased from 2.33 ± 0.54 L/s to 2.46 ± 0.44 L/s and FVC increased from 2.75 ± 0.70 L to 2.84 ± 0.56 L comparing the levels before and after the intervention. The percentage of predicted value of FEV1 (pre-inter- vention 86.7% ± 18.8% vs. post-intervention 93.9% ± 14.5%) and FVC (pre-intervention 89.9% ± 21.8% vs. post-intervention 95.0% ± 16.3%) improved to some degree after the intervention, but all the parameters of lung function assessment did not show significant change. In the 49 participants who had lung function assessed, the prevalence of self-reported eye and respiratory symptoms that occurred in the past 3 months from the post- intervention survey decreased for all symptoms examined compared to the reported prevalence from baseline, including tearing, sore eyes, red eyes, runny nose, nose congestion, continuous sneeze, phlegm, and fever (Table 7).

Among the total study population, a similar reduction in the percent prevalence of all examined symptoms was apparent following the intervention compared to baseline (Table 7). Significant reductions were observed for the prevalence of tearing, which decreased from 5.6% to 0%, sore eyes which decreased from 3.6% to 1.5%, and for red eyes which decreased from 3.9% to 1.0%. Further, there were significant reductions in the percentage of participants reporting phlegm (pre 6.8% vs. post 3.4%; p = 0.01) and fever (pre 5.3% vs. post 1.7%; p < 0.01; Table 7).

DISCUSSION

The results from this intervention suggest that stove improvement and associated behavioral changes related to fuel use, such as opening windows while cooking, may be effective in mitigating the negative health consequences associated with IAP.

Table 3. Selected characteristics of households for questionnaire survey and for IAP monitoring (%).

	Households for questionnaire survey (n = 413)	Households for IAP monitoring (n = 8)
House area (m²)[a]		
<60	14.6	12.5
60 - 79	33.0	25.0
80 - 99	22.3	25.0
100~	30.1	37.5
Biomass use	100.0	100.0
Stoves without smoke door[b]	88.1	75.0
Chimney lower than eave[c]	77.9	87.5

[a]Household areas were classified by quantiles based on the data of questionnaire survey. One record was missed; [b]Two records were missing; [c]Thirty-three records were missing.

Table 4. Comparison of PM_4 levels before and after intervention ($\mu g/m^3$).

Sampling spots	Time	N[a]	Mean	SD	Median	p-value[b]
Kitchen	Before	8	774	756	568	0.08
	After	8	223	104	197	

Bedroom	Before	8	135	149	100	0.20
	After	8	176	28	163	
Total	Before	8	455	621	183	0.46
	After	8	200	77	163	

[a]Sample size of households for IAP monitoring; [b]Wilcoxon signed rank test comparing data before and after intervention.

Table 5. Comparison of CO levels before and after intervention (ppm).

Sampling spots	Time	N[a]	Mean	SD	Median	p-value[b]
Kitchen	Before	8	3.81	3.25	5.26	0.94
	After	8	3.00	1.57	3.05	
Bedroom	Before	8	2.99	2.79	2.31	0.84
	After	8	2.80	1.77	2.53	
Total	Before	8	3.40	2.96	3.81	0.72
	After	8	2.90	1.62	2.86	

[a]Sample size of households for IAP monitoring; [b]Wilcoxon signed rank test comparing data before and after intervention.

Table 6. Changes in lung function before and after intervention (n = 49).

Variables	Observed		Percentage predicted (%)		Mean changes[a]	t[b]	p-value[b]
	Before	After	Before	After			
VC (L)	3.20 ± 0.57	3.10 ± 0.51	104.5 ± 18.3	103.6 ± 16.0	−0.9	0.32	0.75
FVC (L)	2.75 ± 0.70	2.84 ± 0.56	89.9 ± 21.8	95.0 ± 16.3	5.1	1.11	0.27
FEV1 (L/s)	2.33 ± 0.54	2.46 ± 0.44	86.7 ± 18.8	93.9 ± 14.5	7.1	1.91	0.06
FEF 25% - 75% (L/s)	2.87 ± 0.85	2.71 ± 0.56	87.9 ± 26.9	84.5 ± 17.2	−3.5	−1.61	0.12

[a]Mean changes based on percentage predicted value of lung function (the values after intervention were subtracted by the values before intervention); [b]Adjusted using a linear mixed model for cooking years up to date of survey, using open fire, no fans in kitchens, location of kitchen and never opening windows when cooking.

Table 7. Prevalence rates of symptoms before and after intervention among spirometry participants and total subjects (%).

Symptoms	Spirometry participants (n = 49)			Total subjects (n = 413)		
	Before	After	p-value[a]	Before	After	p-value[b]
Tearing	14.3	2.0	0.01	5.6	0	–
Sore eyes	8.2	4.1	0.41	3.6	1.5	0.04
Red eyes	12.2	2.0	0.06	3.9	1.0	0.01
Runny nose	2.0	0	-	2.2	1.5	0.70
Nose congestion	4.1	0	-	2.7	1.0	0.11
Continuous sneeze	6.1	0	-	2.4	1.2	0.34
Phlegm	8.2	4.1	0.41	6.8	3.4	0.01
Fever	10.2	2.0	0.10	5.3	1.7	<0.01

[a]McNemar's Test; [b]Adjusted for age, cooking years up to date of survey, using open fire, location of kitchen and never open windows when cooking using logistic regression.

Specifically, we observed a reduction in average PM_4 levels, particularly in the kitchen area, and in the prevalence of symptoms associated with IAP following implementation of the stove improvement and behavioral intervention. Further, FEV1 favorably increased following the intervention compared to baseline, although this effect was not statistically significant. The households selected for IAP assessment were shown to be representative of the larger study population with respect to home area, fuel type used, and stove features and ventilation practices, suggesting that our findings may be generalizable to the overall study population. While the average bedroom concentration of PM_4 was slightly higher after stove improvement, this is potentially attributed to the small sample size of monitored households and/or the limited duration of each measurement period (i.e., 24 hours).

Since few nations have indoor air quality standards for PM_4, we can only compare the result with the limit for PM_{10} in domestic air. In China, the national standard for indoor air quality dictates that the average concentration of PM_{10} should not exceed 150 µg/m³ [19]. While stove improvement was demonstrated to reduce PM_4 levels, the concentration of PM_4 in indoor air was still higher than this standard for PM_{10}. Using solid fuels was the root of IAP in rural households for the limit of economic status. It would be difficult to reduce the indoor air pollutants to very low levels unless clean fuels were prevalent in rural households. While our results indicated significant behavioral modifications concerning fuel use habits, some lifestyles, such as using portable open fires to heat tea, were still popular after the intervention. Consequently, this may have contributed to the relatively high levels of PM_4 even after stove improvement in some households.

Changes in CO levels generally suggested a mild trend of decline, but the measured concentrations were lower than expected. Specifically, 24 hour average concentrations of CO before and after intervention were consistently below current standards and guidelines (i.e., WHO guideline values of 5.6 ppm/7 mg/m³ for 24 h exposures), and in some cases CO concentrations were close to the detection limits of the diffusion tubes. Since these were 24-h concentrations, concentrations may have been higher during cooking or when doors and windows were closed at night, but this was not observable in our data [16]. Furthermore, about 83% of the surveyed women use to open windows in the winter and 84% opened kitchen windows for ventilation when cooking even before the behavior intervention; therefore, generated CO may have diffused to others rooms or outdoors. For both CO and PM_4, the percent reduction in levels before and after the intervention in the field study were lower compared to previously conducted controlled tests [16]. This may primarily be due to other factors associated with levels of these pollutants, such as individual cooking practices and ventilation patterns, and also due to the fact that the controlled tests were conducted only during periods when the stove was burning.

In our study, socioeconomic differences, individual biological variations, and some potential confounders that might affect lung function among populations were controlled because we compared the changes in the same people before and after intervention. In addition, we transformed the observed values to percentages of predicted values in order to control for age and possible changes of body composition during the 2-year study period, and used linear mixed models to assess the improvement of lung function parameters, adjusting for other potential confounders. In some follow-up studies on lung function, spirometry results may have been affected by participants who repeated the test several times and developed experience and skills to perform better on the respiratory function test [20]. However, this was unlikely to have influenced the results in our study given the relatively long interval between the first and second lung function assessments. Data in this study suggested the possible improvement of FEV1 in women to some degree after IAP mitigation. However, this finding should be interpreted cautiously since other parameters of lung function did not show similar improvement. The post-interven- tion examination of lung function was conducted roughly 5 months after the large-scale stove intervention. It would be unlikely for these parameters to improve right after intervention measures had been taken. Follow-up studies are warranted in order to evaluate the long term effect of IAP mitigation on lung function.

On the other hand, we observed significant decreases in the prevalence rates of some eye and respiratory symptoms, as well as for fever, following the intervention. In other community studies carried out in developing countries, investigators also found that biomass exposure was consistently associated with chronic respiratory symptoms although effects on lung function were variable or small [5] [20] [21]. These studies suggested that symptoms were more sensitive to variation of IAP levels. Since accurate disease records were not available in this region, the symptoms survey was alternatively used to measure the variation in health status. Given that eye and respiratory symptoms reflect stimulation of the mucosas and aggravation or attack of diseases, the decreases in prevalence of these symptoms provide a potential indication of improvement in health status following the intervention and reduction in IAP levels. Our results also indicated similarly decreasing trends in the prevalence of symptoms in the 49 women who took part in the lung function test, which suggested health improvement after intervention similar to the overall study population. These women also had relatively higher prevalence rates of all symptoms except runny nose during the baseline period and more significant reduction in the post-interven- tion interview. This might be attributed to the fact that people in poorer health condition were more likely to participate in the IAP mitigation activities and would get more improvement in health. However, we note that the study personnel and participants in the study were not blinded to the intervention status, and therefore there is a possibility of biased reporting for the subjective physical symptoms. In addition, it is possible that some reported symptoms may have been due to other causes aside from the effects of IAP, although participants were asked about the prevalence of symptoms over a three month period before the interview, which provided some determination of whether the symptoms were chronic in nature rather than from acute respiratory infections.

There were some additional limitations in this study. First, combustion of solid fuels was the source of indoor air pollutants in rural regions of China. However, it would not be feasible to implement cleaner energy (e.g. natural gas and electricity) given the poor economic conditions among the rural households and undeveloped infrastructure. The primary goal of this intervention was therefore to inform the public of the adverse effects of IAP from biomass and assist them in replacing traditional stoves. The appearance, usage, and maintenance of the new stoves were similar to the traditional ones and thus the effect of the intervention is likely to be sustainable. Second, the exposure assessment conducted in our study utilized area sampling for a relatively short time period (i.e., 24 hours) rather than using personal monitors for the study subjects. This approach provided a general indication of the levels of IAP in the monitored households, but might not capture the variability in exposures and could either underestimate or overestimate personal exposure depending on the placement of the monitors in relation to the participants' usual activities. Third, our study was relatively small in terms of the number of monitored households and subjects included in lung function testing, and the timeline of the project required a shorter than ideal interval between the baseline and follow-up periods and therefore follow-up studies will be needed. Finally, some risk behaviors practiced for generations were difficult to change in the one- or two-year intervention period.

CONCLUSION

In conclusion, we have shown that an intervention consisting of stove improvement and behavioral modifications related to fuel use may be an effective and viable method to mitigate the deleterious health effects of IAP in developing regions. Specifically, our results suggest that such an intervention may reduce levels of PM_4 and physical symptoms associated with IAP exposure, and in addition may improve some measures of lung function, although the effects reported in our study are not statistically significant due to the relatively small size of our study and the symptoms assessed are based on self-report. Follow-up studies with a longer assessment period between the pre and post intervention periods are clearly needed to properly evaluate changes in lung function associated with these stove improvements, as well as the long-term effects of the intervention.

ACKNOWLEDGEMENTS

The authors would like to express special gratitude to all the study subjects in the study and the personnel in the area of Gansu Province, China, which have supported the study. Funding for this work was provided by the World Bank through financial support from the Energy Sector Management Assistance Program (ESMAP), Department for International Development, UK (DFID) and Swedish International Development Agency (SIDA) and also partly supported by Fogarty training grants D43TW 008323 and D43TW 007864-01 from the National Institutes of Health.

REFERENCES

1. World Health Organization (2013) World Health Statistics.http://www. who.int/gho/publications/world_health_statistics/2013/en/

2. World Health Organization (2007) Indoor Air Pollution: National Burden of Disease Estimates. World Health Organization, Geneva.http://www. who.int/indoorair/publications/indoor_air_national_burden_estimate_ revised.pdf

3. Lim, S., Vos, T., Flaxman, A.D., Danaei, G., Shibuya, K., Adair-Rohani, H., et al. (2012) A Comparative Risk Assessment of Burden of Disease and Injury Attributable to 67 Risk Factors and Risk Factor Clusters in 21 Regions, 1990-2010: A Systematic Analysis for the Global Burden of Disease Study 2010. The Lancet, 380, 2224-2260.http://dx.doi. org/10.1016/S0140-6736(12)61766-8

4. Kurmi, O.P., Semple, S., Simkhada, P., Smith, W.C. and Ayres, J.G. (2010) COPD and Chronic Bronchitis Risk of Indoor Air Pollution from Solid Fuel: A Systematic Review and Meta-Analysis. Thorax, 65, 221- 228. http://dx.doi.org/10.1136/thx.2009.124644

5. Regalado, J., Perez-Padilla, R., Sansores, R., Paramo Ramirez, J.I., Brauer, M., Pare, P., et al. (2006) The Effect of Biomass Burning on Respiratory Symptoms and Lung Function in Rural Mexican Women. American Journal of Respiratory and Critical Care Medicine, 174, 901- 905. http://dx.doi.org/10.1164/rccm.200503-479OC

6. Díaz, E., Smith-Sivertsen, T., Pope, D., Lie, R.T., Díaz, A., McCracken, J., et al. (2007) Eye Discomfort, Headache and Back Pain among Mayan Guatemalan Women Taking Part in a Randomised Stove Intervention Trial. Journal of Epidemiology and Community Health, 61, 74-79. http:// dx.doi.org/10.1136/jech.2006.043133

7. Liu, S., Zhou, Y., Wang, X., Wang, D., Lu, J., Zheng, J., et al. (2007) Biomass Fuels Are the Probable Risk Factor for Chronic Obstructive Pulmonary Disease in Rural South China. Thorax, 62, 889-897. http:// dx.doi.org/10.1136/thx.2006.061457

8. Mishra, V. (2003) Effect of Indoor Air Pollution from Biomass Combustion on Prevalence of Asthma in the Elderly. Environmental Health Perspectives, 111, 71-78.http://dx.doi.org/10.1289/ehp.5559

9. Mestl, H.E., Aunan, K., Seip, H.M., Wang, S., Zhao, Y. and Zhang, D. (2007) Urban and Rural Exposure to Indoor Air Pollution from Domestic Biomass and Coal Burning across China. Science of the Total Environment, 377, 12-26.http://dx.doi.org/10.1016/j.scitotenv.2007.01.087f

10. World Health Organization (2009) Global Health Risks: Mortality and Burden of Disease Attributable to Selected Major Risks.http://www.who. int/healthinfo/global_burden_disease/GlobalHealthRisks_report_full. pdf

11. Jin, Y., Ma, X., Chen, X., Cheng, Y., Baris, E. and Ezzati, M. (2006) Exposure to Indoor Air Pollution from Household Energy Use in Rural China: The Interactions of Technology, Behavior, and Knowledge in Health Risk Management. Social Science Medicine, 62, 3161-3176. http://dx.doi.org/10.1016/j.socscimed.2005.11.029

12. He, G., Ying, B., Liu, J., Gao, S., Shen, S., Balakrishnan, K., et al. (2005) Patterns of Household Concentrations of Multiple Indoor Air Pollutants in China. Environmental Science Technology, 39, 991-998. http://dx.doi. org/10.1021/es049731f

13. Chapman, R.S., He, X., Blair, A.E. and Lan, Q. (2005) Improvement in Household Stoves and Risk of Chronic Obstructive Pulmonary Disease in Xuanwei, China: Retrospective Cohort Study. British Medical Journal, 331, 1050.http://dx.doi.org/10.1136/bmj.38628.676088.55

14. Lan, Q., Chapman, R.S., Schreinemachers, D.M., Tian, L. and He, X. (2002) Household Stove Improvement and Risk of Lung Cancer in Xuanwei, China. Journal of the National Cancer Institute, 94, 826-835. http://dx.doi.org/10.1093/jnci/94.11.826

15. Shen, M., Chapman, R.S., Vermeulen, R., Tian, L., Zheng, T., Chen, B.E., et al. (2009) Coal Use, Stove Improvement, and Adult Pneumonia Mortality in Xuanwei, China: A Retrospective Cohort Study. Environmental Health Perspectives, 117, 261-266.http://dx.doi.org/10.1289/ehp.11521

16. Zhou, Z., Jin, Y., Liu, F., Cheng, Y., Liu, J., Kang, J., et al. (2006) Community Effectiveness of Stove and Health Education Interventions for Reducing Exposure to Indoor Air Pollution from Solid Fuels in Four Chinese Provinces. Environmental Research Letters, 1, 1-12. http:// dx.doi.org/10.1088/1748-9326/1/1/014010

17. Jin, Y., Zhou, Z., He, G., Wei, H., Liu, J., Liu, F., et al. (2005) Geographical, Spatial, and Temporal Distributions of Multiple Indoor Air Pollutants in Four Chinese Provinces. Environmental Science Technology, 39, 9431-9439. http://dx.doi.org/10.1021/es0507517

18. Mu, K. and Liu, S. (1990) National Compilation of Normal Value of Lung Function. Peking Union Medical College, Beijing Medical University Joint Publishing House, Beijing.

19. Ministry of Health (2002) Indoor Air Quality Standard Ministry of Health, China.http://www.moh.gov.cn/publicfiles//business/htmlfiles/ zwgkzt/pwsbz/index.htm

20. Smith-Sivertsen, T., Dıaz, E., Pope, D., Lie, R.T., Dıaz, A., McCracken, J., et al. (2009) Effect of Reducing Indoor Air Pollution on Women's Respiratory Symptoms and Lung Function: The RESPIRE Randomized Trial, Guatemala. American Journal of Epidemiology, 170, 211-220. http://dx.doi.org/10.1093/aje/kwp100

21. Rinne, S., Rodas, E., Bender, B., Rinne, M., Simpson, J., Galer-Unti, R., et al. (2005) Relationship of Pulmonary Function among Women and Children to Indoor Air Pollution from Biomass Use in Rural Ecuador. Respiratory Medicine, 100, 1208-1215.http://dx.doi.org/10.1016/j. rmed.2005.10.020

Chapter 8

SPATIAL ANALYSIS ON THE CONCENTRA-TIONS OF AIR POLLUTANTS IN BASRA PROVINCE (SOUTHERN IRAQ)

Shukri I. Al-Hassen[1], Abdul Wahab A. Sultan[2], Adnan A. Ateek[2], Hamid T. Al-Saad[3], Salah Mahdi[3], Abdulzahra A. Alhello[3]

[1]Department of Geography, University of Basra, Basra, Iraq
[2]Technical College, Southern Technical University, Basra, Iraq
[3]Department of Environmental Chemistry, University of Basra, Basra, Iraq

ABSTRACT

This paper aims to analyze the geographic distribution of air pollutant concentrations in Basra Province, Southern Iraq, and to cartographically determine the spatial variation of air pollution levels as well as to recognize the hottest spots of air pollution within the study area, and conclude that the levels of air pollution in the study area are spatially varied, with an irregular spatial pattern and some hotspots.

INTRODUCTION

Air pollution may be defined as any atmospheric condition in which certain substances are present in such concentrations that they can produce harmful effects on man and his environment. An air pollutant, however, is any gas or substance (such as SO_x, NO_x, CO, and HCs) or particulate matter (such as smoke, dust, fumes, and aerosols) that leads to ambient air contamination. A pollutant may originate from natural or anthropogenic sources, or both. Pollutants occur throughout much of the troposphere; however, pollution close to the earth's surface within the boundary layer is of most concern because of the relatively high concentrations resulting from sources at the surface [1] [2].

Air pollutant concentrations depend mainly on the total mass of pollution emitted into the atmosphere, together with the atmospheric conditions that affect its fate and transport. Obviously, air pollution has many and varied sources, including cars, smokestacks, and other industrial inputs into the atmosphere as well as wind erosion of soil. Large emissions from both anthropogenic and natural sources over long periods enhance concentrations, as do the chemical and physical properties of these pollutants. For example, when nitrogen oxides and hydrocarbons in car exhaust are emitted into warm, sunlit air, they readily form ozone molecules (O_3). Similarly, the solubility of a pollutant affects how efficiently it is removed by rainfall. In addition, atmospheric conditions have a major effect upon pollutants once these pollutants are emitted into (e.g., nitrogen oxides from car exhaust) or formed within (e.g., O_3) the atmosphere. Pollution dispersal is controlled by atmospheric motion, which is affected by wind, stability, and the vertical temperature variation within the boundary layer. Stability, in turn, influences both air turbulence and the depth at which mixing of polluted air takes place ([2] ; see also [3] -[7]).

In the study area (Basra province), located in southern Iraq, air pollution is of major public concern, which is currently the object of extensive scientific research. Studies of Al-Asadi [8], Al-Mayahi [9], Al-Imarah et al. [10], Al-Saad et al. [11], Garabedian [12], Al-Hassen [13], Qassim [14], Douabul et al. [15], Sultan et al. [16], Karmalla et al. [17], Abdullah and Hussien [18], and Al-Hassen et al. [19], are an example of some local-scale research in this respect.

The study area is rich in petroleum and many existing industrial and human activities as the main sources of gaseous emissions around it (Figure 1). These are emission sources to contaminate the ambient air in Basra. According to above mentioned studies, Basra was recorded, in the last two decades, high quantities of gaseous emissions causing elevated levels of outdoor air pollution, and that concentration of some air pollutants was at risk to the public health. One of the most important driving forces is atmospheric conditions in this region.

(a)

(b)

(c)

(d)

Figure 1. Photography of some gaseous emission sources in the study area. (a) Petrochemical plant; (b) Najybia power station; (c) Emissions of petroleum exploitation near Burchesya; (d) Iranian Abadan refinery offshore seeba.

The previous studies, however, are focused on values of gaseous pollutant concentration and its implications on the human health in the study area; thus the objective of the present study is to spatially analyze the geographic distribution of air pollutant concentrations in Basra, and to cartographically determine the spatial variation of air pollution levels on the map of this region. The major aim is to recognize the hottest spot of air pollution within whole areas of Basra province. Basra has a hot desert climate (Köppen climate classification BWh), like the rest of the surrounding region, though it receives slightly more precipitation than inland locations due to its location near the coast. During the summer months, from June to August, Basra is consistently one of the hottest cities on the planet, with temperatures regularly exceeding 40°C (104°F) and approaching 45°C (113°F) in July. In winter Basra experiences mild weather with average high temperatures around 20°C (68°F). On some winter nights, minimum temperatures are below 0°C (32°F). High humidity—sometimes exceeding 90%—is common due to the proximity to the marshy Persian Gulf [20].

MATERIALS AND METHODS

As shown in Figure 2, seventeen sampling stations were chosen in the study area. The selected stations divided to cover the eastern and western region within the study area, and the geographic distribution of sampling stations was taken in consideration in the vicinity of human settlements and involved a variety of local environments (see Table 1).

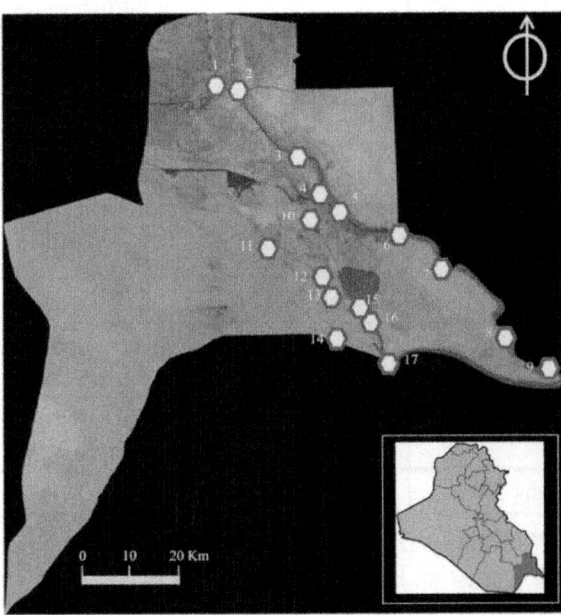

Figure 2. Map of the study area (Basra Province, Southern Iraq), showing the selected sampling stations. Note: Numbers of sampling stations correspondences their names listed in Table 1 and Table 2.

Table 1. Type of environment in the selected sampling stations, and the major sources of gaseous emissions in the study area.

No.	Sampling station	Type of environment	Major source of air pollution
1	Midaina	Agricultural/industrial/urban	Petroleum exploitation and urban contaminates
2	Qurna	Agricultural/industrial/urban	Petroleum exploitation and urban contaminates
3	Dayer	Agricultural/industrial/urban	Petroleum exploitation and urban contaminates
4	Garmatt Ali	Agricultural/urban	Power station
5	Ashar	Urban	Urban contaminates
6	Abo Khaseeb	Agricultural/urban	Urban contaminates
7	Seeba	Agricultural/industrial	Petroleum industry and exploitation
8	Faw	Agricultural/urban	Urban contaminates
9	Ras Absha	Marine	Natural contaminates
10	Shatt Al-Basra	Riverine	Natural and urban contaminates
11	Burchesya	Industrial	Petroleum industry and exploitation
12	Petrochemical plant area	Industrial	Petroleum industry
13	South gas plant area	Industrial	Petroleum industry
14	Safwan	Urban	Urban contaminates
15	Khor Al-Zubayer	Industrial/urban	Industrial and urban contaminates
16	Gas terminal	Industrial/marine	Petroleum industry and shipping
17	Umm Qasr	Urban/marine	Urban contaminates

The selected stations were in order to monitor the concentrations of gaseous pollutants released into ambient air of the study area. A variety of gaseous pollutants such as carbon monoxide (CO), carbon dioxide (CO_2), sulfate oxides (SO_x), nitrogen oxides (NO_x), ozone (O_3), petroleum hydrocarbons (HCs), methane (CH_4), hydrogen sulfide (H_2S), and formaldehyde (HCHO), measured in this work. Concentrations of HCs, NO_x, SO_x, HCHO, O_3, and CO_2 measured using the portable detection instrument of Drager Chip-Measurement System, Germany, whereas the portable instrument of RK1 Gas Monitoring Eagle II, USA, detected the pollutants of CH_4, H_2S, and CO.

Fieldwork carried out during a one daytime of spring 2015, and all measurements were done at the same time in purpose of obtaining on readings approaching to reality. Therefore, the working team divided into two groups, one of them work in the eastern region (Midaina, Qurna, Dayer, Garmatt Ali, Ashar, Abo Khaseeb, Seeba, Faw, and Ras Absha), while the other group work in the western region (Shatt Al-Basra, Burchesya, Petrochemical plant area, South Gas Plant area, South Gas Plant area, Safwan, Khor Al-Zubayer, Gas terminal, Umm Qasr). The procedure of measuring was as described by Douabul et al. [15].

RESULTS AND DISCUSSION

Table 2 lists the obtained results for concentrations of some air pollutants in the given study area and period. These concentrations recorded from the direct readings displayed on the screen of both employed detectors. The values were adjusted in a statistic form to make a more geographic mode. Thus, the values have been graphically and cartographically represented in Figure 3 and Figure 4, respectively. In this study, data analysis and explanation will conducted in the terms of each given element with the emphasis on spatial variation and geographic distribution of air pollution, as follows:

- Carbon monoxide (CO) is a colorless, odorless, and tasteless gas that is slightly less dense than air. It is toxic to humans when encountered in concentrations above about 35 ppm. In the atmosphere, it is spatially variable and short lived, having a role in the formation of ground-level ozone [21].

Table 2. Concentrations of air pollutants measured in the study area based on the selected sampling stations, 2015.

No.	Sampling station	CO ppm	CO_2 ppm	NO_x ppm	SO_x ppm	H_2S ppm	HCs ppm	CH_4 ppm	HCHO ppm	O_3ppb	Index*
1	Midaina	4.16	215.12	0.72	0.54	1.53	1.62	8.52	0.42	0.02	25.85
2	Qurna	6.25	286.45	0.83	0.64	1.82	1.93	8.86	0.35	0.03	34.12
3	Dayer	8.92	260.21	0.86	0.63	1.92	3.25	10.21	0.46	0.02	31.83
4	Garmatt Ali	10.23	280.38	0.95	1.25	1.98	5.28	9.52	0.63	0.06	34.47
5	Ashar	12.32	225.32	0.65	0.92	1.21	12.21	13.28	0.92	0.12	29.66

6	Abo Khaseeb	20.63	250.12	0.83	1.28	1.26	24.28	14.25	1.23	0.14	34.89
7	Seeba	30.23	280.11	1.45	2.28	2.68	31.23	25.68	1.86	0.23	41.63
8	Faw	10.24	240.61	0.79	1.68	2.06	22.52	13.34	0.66	0.06	32.44
9	Ras Absha	2.52	180.32	0.42	0.43	1.25	1.74	5.54	0.24	0.01	21.38
10	Shatt Al-Basra	10.68	210.11	0.65	0.72	1.12	7.25	10.11	0.72	0.04	26.82
11	Burchesya	40.23	310.27	4.25	10.23	6.24	30.21	22.65	1.52	0.12	69.30
12	Petrochemical plant area	16.23	200.10	1.31	1.65	1.32	10.53	13.21	0.31	0.06	27.19
13	South gas plant area	20.53	220.31	1.45	3.21	3.1	18.23	10.25	0.62	0.16	30.87
14	Safwan	18.22	210.53	0.93	1.24	2.5	10.82	9.34	0.52	0.13	28.24
15	Khor Al-Zubayer	14.28	226.3	1.23	2.52	1.8	11.23	9.93	0.68	0.09	29.78
16	Gas terminal	18.34	228.23	1.86	4.38	2.1	22.31	16.83	0.84	0.11	32.77
17	Umm Qasr	16.12	180.98	0.98	3.49	3.5	24.63	12.46	0.75	0.07	26.99
	Mean	15.30	235.61	1.15	2.18	2.19	14.01	12.58	0.74	0.08	31.49

Data based on Fieldwork. *Index means a sum of values for the selected parameters divided by its number. It may be expressed, in the other meaning, the intensity of pollution.

As listed in Table 2 and Figure 3 & Figure 4(a), the Burchesya sampling station records the highest value of CO is 40.23 ppm, while the lowest is 2.52 ppm in Ras Abasha station. The mean concentration of CO is 15.30 ppm within the all sampling stations.

- Carbon dioxide (CO_2) is a colorless, odorless gas vital to life on Earth. Carbon dioxide exists in the Earth's atmosphere as a trace gas at a concentration of about 0.04 percent (400 ppm) by volume. It is present in deposits of petroleum oil and natural gas [22].

Table 2 and Figure 3 & Figure 4(b) report that the maximum concentration of CO_2 is 310.27 ppm recorded in the Burchesya sampling station, while the minimum concentration is 180.32 ppm in the Ras Abasha station. In general, the sampling stations in the eastern region of the study area registers values higher than those that in the western stations. The mean concentration of CO_2 is 235.61 ppm.

- Nitrogen oxide (NO_x) is a prominent air pollutant; it may refer to a binary compound of oxygen and nitrogen, or a mixture of such compounds. This reddish-brown toxic gas has a characteristic sharp, biting odor and is a prominent air pollutant [23].

Table 2 and Figure 3 & Figure 4(c) indicate that the highest concentration of NO_x registered in the study area was 40.23 at the Burchesya sampling station, this may be a record value to compare with given in the previous studies yet. The lowest concentration was 0.42 in the Ras Abasha station. The mean concentration of NO_x is 1.15 ppm.

- Sulfur oxide (SO_x) refers to many types of sulfur and oxygen containing compounds such as SO, SO_2, SO_3, S_7O_2, S_6O_2, S_2O_2, etc. SO_x is a toxic gas with a pungent, irritating, and rotten smell [24].

In this study, as shown in Table 2 and Figure 3 & Figure 4(d), the maximum concentration of SO_x was 10.23 ppm in the Burchesya sampling station, while the minimum concentration is 0.42 ppm in the Ras Abasha station. The mean concentration of SO_x is 2.18 ppm.

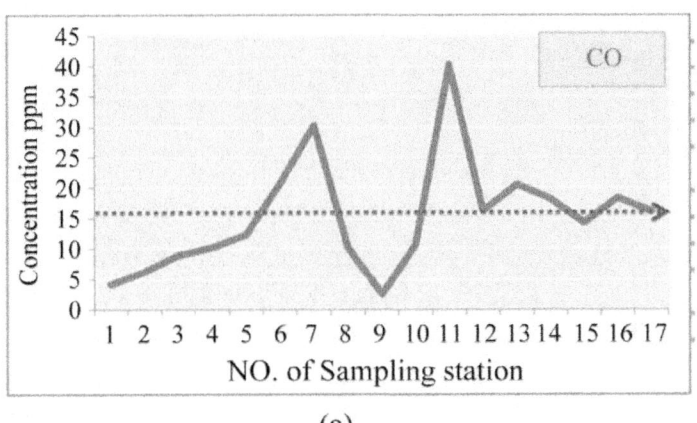

(a)

(b)

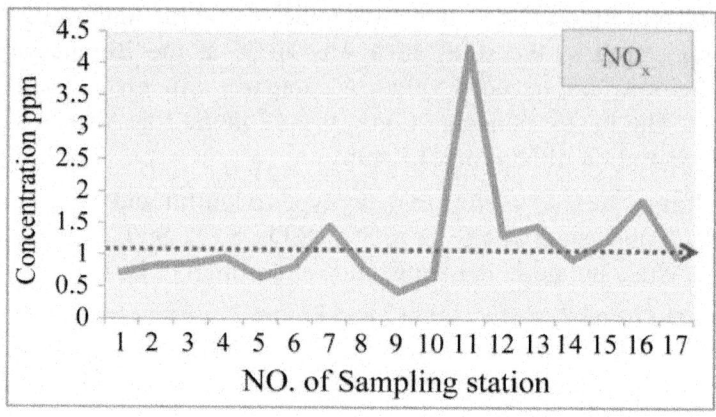

(c)

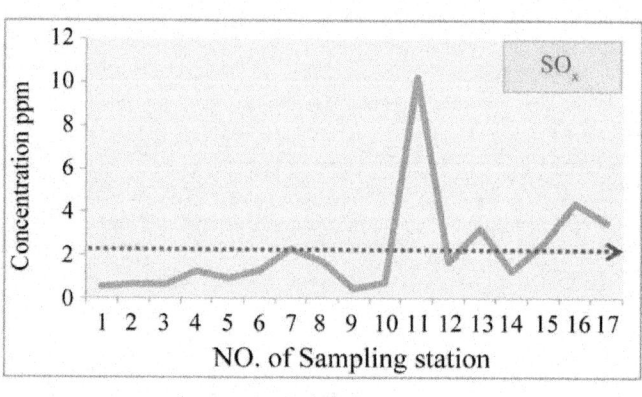

(d)

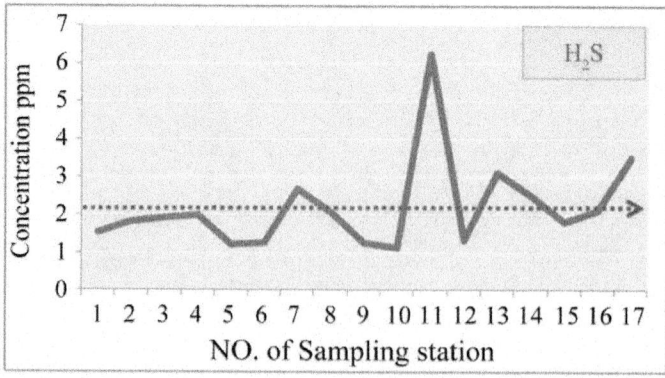

(e)

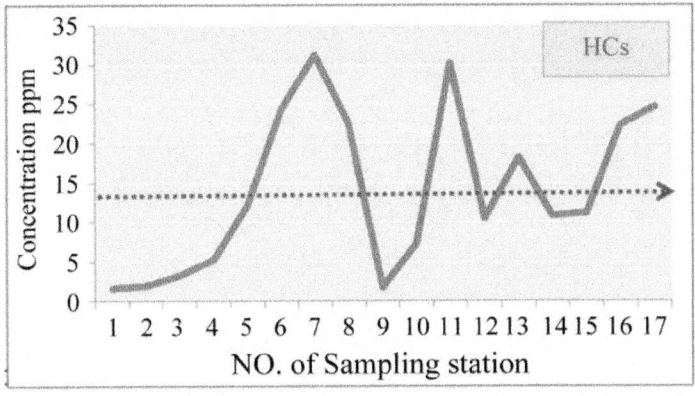

(f)

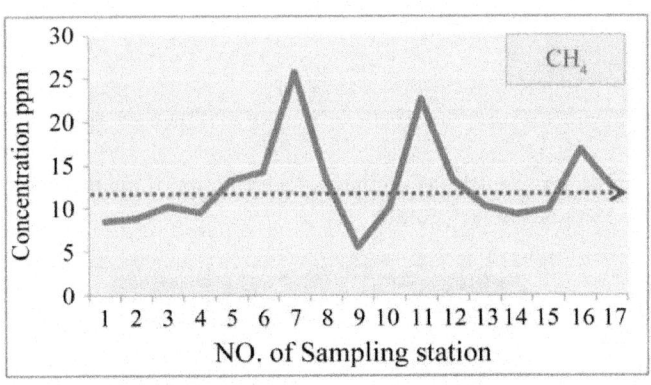

(g)

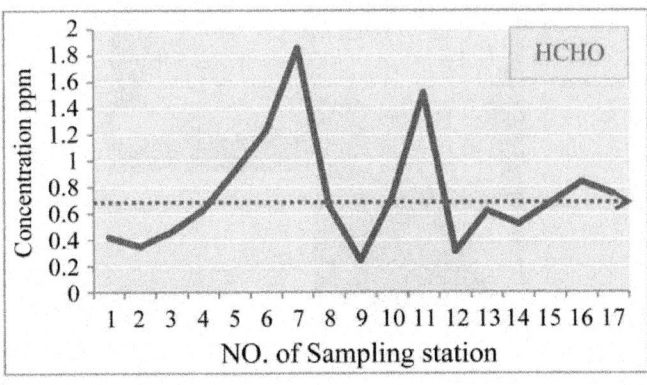

(h)

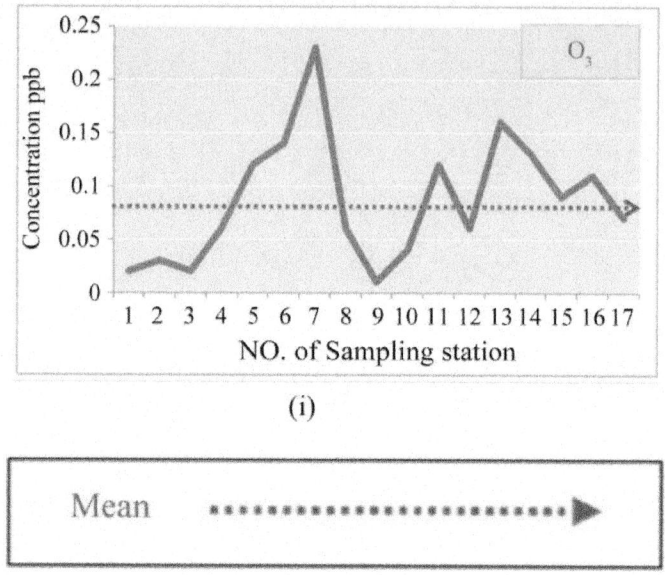

(i)

Figure 3. Graphic representation of the concentrations of air pollutants measured in the study area. Data based on Table 2.

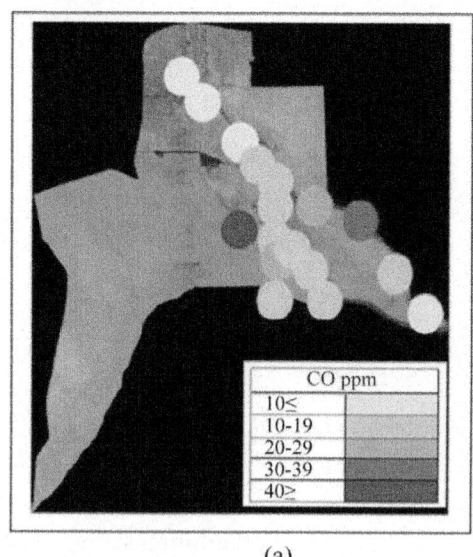

(a)

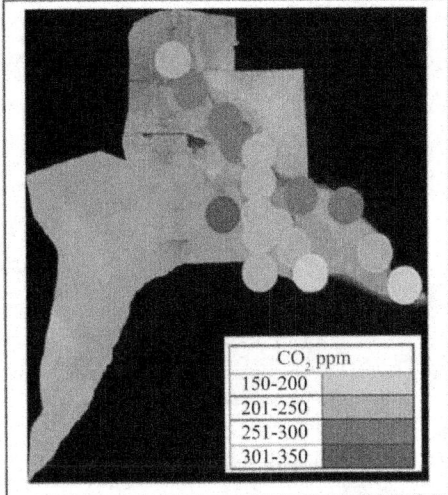

(b)

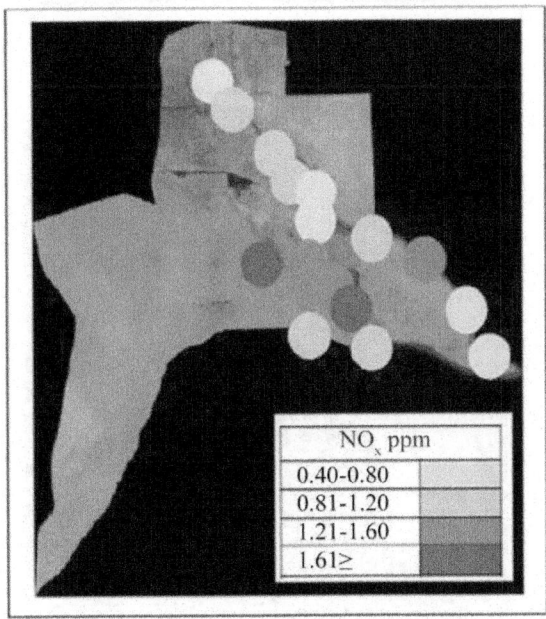

(c)

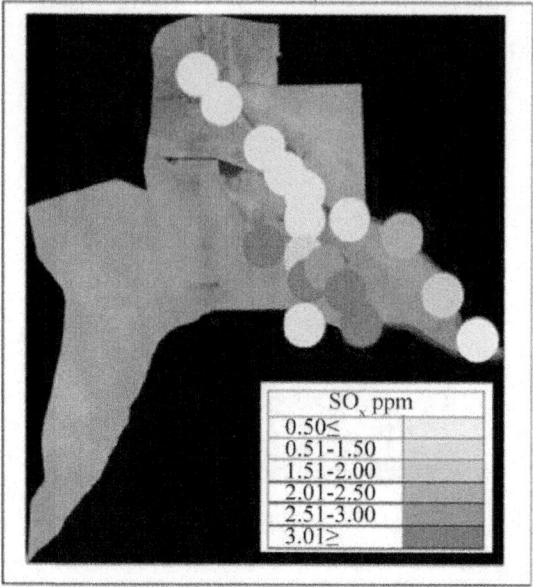

(d)

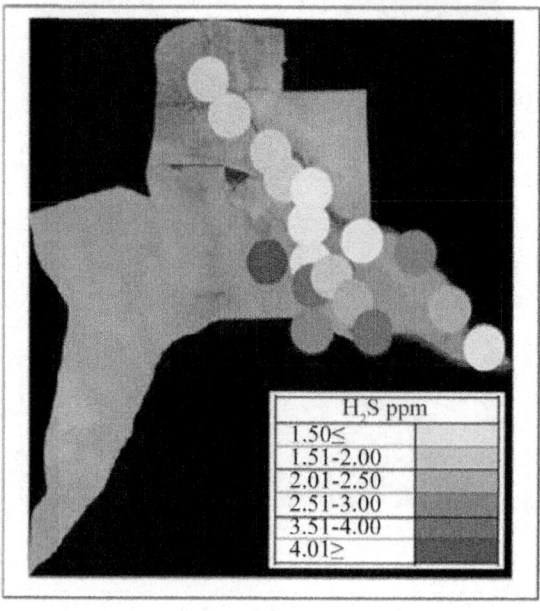

(e)

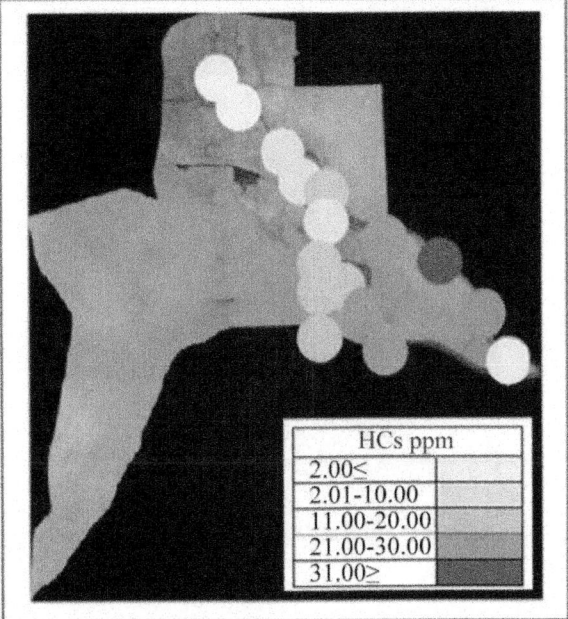

HCs ppm	
2.00≤	
2.01-10.00	
11.00-20.00	
21.00-30.00	
31.00>	

(f)

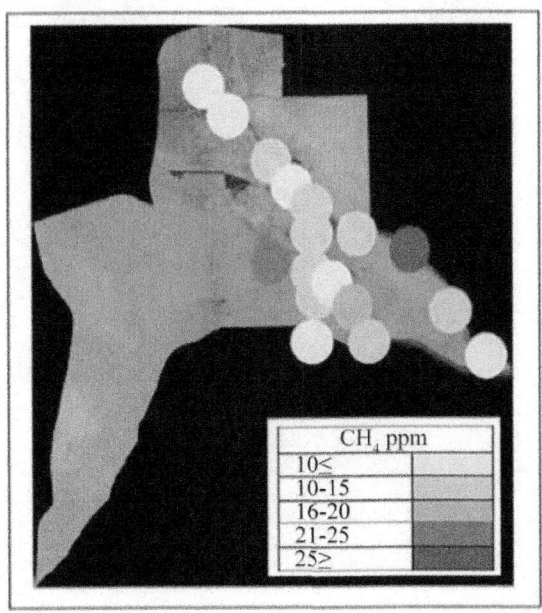

CH₄ ppm	
10≤	
10-15	
16-20	
21-25	
25>	

(g)

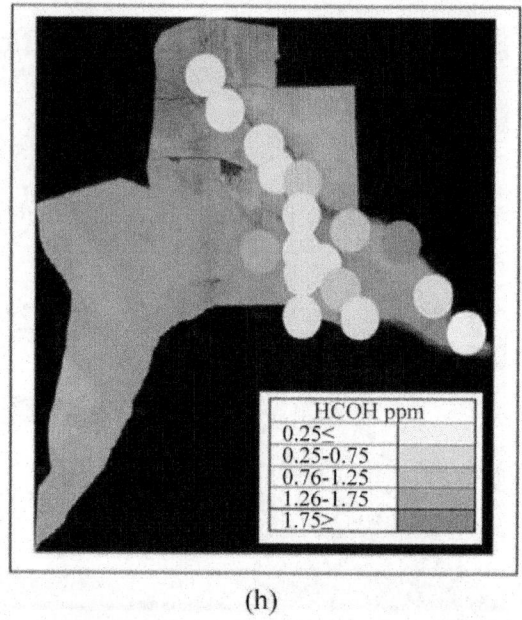

HCOH ppm	
0.25<	
0.25-0.75	
0.76-1.25	
1.26-1.75	
1.75>	

(h)

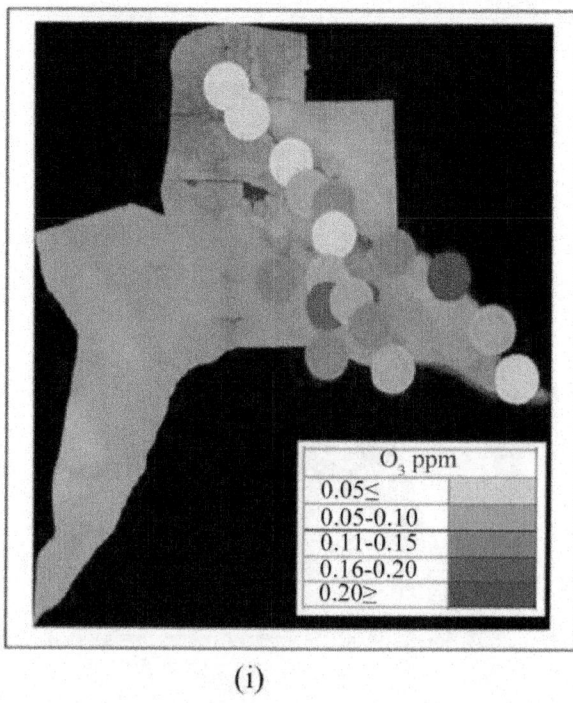

O_3 ppm	
0.05≤	
0.05-0.10	
0.11-0.15	
0.16-0.20	
0.20≥	

(i)

Figure 4. Cartographic representation of the concentrations of air pollutants measured in the study area. Data based on Table 1.

- Hydrogen sulfide is the chemical compound with the formula (H_2S). It is a colorless gas with the characteristic foul odor of rotten eggs; it is heavier than air, very poisonous, corrosive, flammable, and explosive [25].

Table 2 and Figure 3 & Figure 4(e) refer to the highest value of H_2S recorded in this study was 6.24 ppm in the Burchesya sampling station, while the lowest value was 1.21 ppm in the Ashar station. The mean concentration is 2.19 ppm.

- Hydrocarbon (HCs) is an organic compound consisting entirely of hydrogen and carbon. Hydrocarbon poisoning such as that of benzene and petroleum usually occurs accidentally by inhalation or ingestion of these cytotoxic chemical compounds [22].

HCs gas concentrations in the study area were high, in general. The maximum concentration is 31.23 ppm registered at the Seeba sampling station, whereas the minimum concentration is 1.62 in the Midaina station. The mean concentration of HCs is 14.01 ppm, as shown in Table 2 andFigure 3 & Figure 4(f).

- Methane is a chemical compound with the chemical formula (CH_4). It is the simplest alkane and the main component of natural gas. Methane is not toxic, yet it is extremely flammable and may form explosive mixtures with air [26].

The highest values of CH_4, as indicated in Table 2 and Figure 3 & Figure 4(g), were concentered in places with the petroleum industry and exploitation within the study area. The maximum concentration, however, is 25.68 ppm in the Seeba station, while the minimum concentration is 5.54 ppm in the Ras Abasha station. The mean concentration of CH_4 is 12.58 ppm.

- Formaldehyde (HCHO) is a colorless, highly toxic, and flammable gas at room temperature that is slightly heavier than air. It has a pungent, highly irritating odor that is detectable at low concentrations, but may not provide adequate warning of hazardous concentrations for sensitized persons [27].

Table 2 and Figure 3 & Figure 4(h) show that the concentration of HCHO records the maximum value is 1.86 ppm in the Seeba station, while the minimum value is 0.24 ppm recorded in the Ras Abasha station. The mean concentration of HCHO is 0.74 ppm.

- Ozone is an inorganic molecule with the chemical formula (O_3). It is a pale blue gas with a distinctively pungent smell. This same high oxidizing potential, however, causes ozone to damage mucous and respiratory tissues in animals, and also tissues in plants, above concentrations of about 100 ppb. This makes ozone a potent respiratory hazard and pollutant near ground level [28].

In the study area, as shown in Table 2 and Figure 3 & Figure 4(i), the concentrations of ground level ozone (O_3) were largely varied within the selected sampling stations. The highest value is 0.23 ppb in the Seeba station, whereas the Ras Abasha station register the lowest value is 0.01 ppb. Thus, the mean concentration of O_3 is 0.08 ppb.

In general, the concentrations of air pollutants registered in the study area were spatially varied, it seems somehow a random pattern of distribution in the terms of each pollutant (see Figure 4). However, an overall spatial pattern may be drawn by using the index of pollution showing in Figure 5 based on Table 2. This index is a result of summing all values of each element dividing by its numbers, the resulting value is an approximate indicator of the spatial concentration of a pollutant. To simplify explaining the causes of spatial variation in pollutant concentrations, the mentioned Table 1 lists the major gaseous emission sources affecting air quality in the study area.

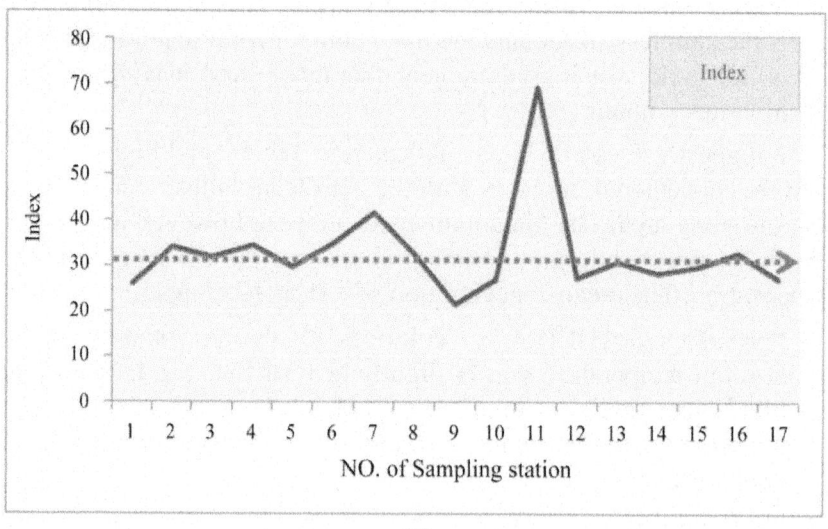

(a)

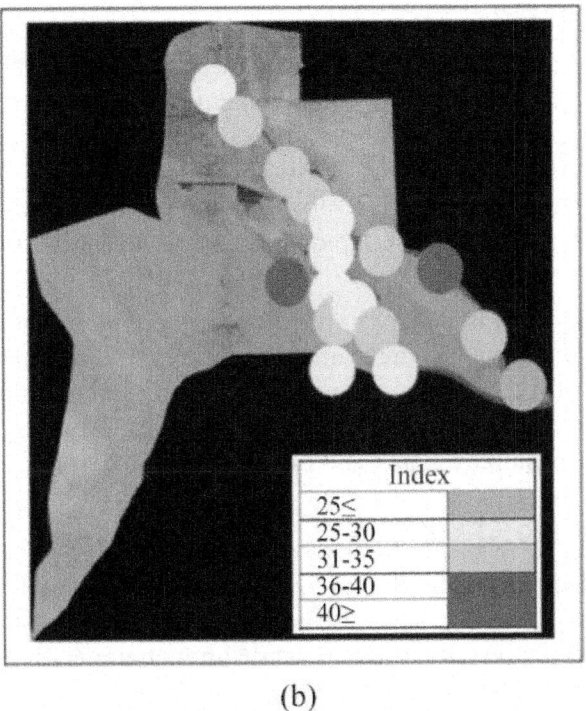

Index

25<	
25-30	
31-35	
36-40	
40≥	

(b)

Figure 5. Geographical distribution of the levels of air pollution in the study area according to index of pollution. Data based on Table 2. Data based on Table 1.

As shown in Figure 5(a) & Figure 5(b), the elevated levels of air pollution focuses on the western region of the study area, this is due to existing industrial pollution sources. The sampling stations of Qurna, Dayer, and Garmatt Ali also report a significant index values with 34.12, 31.83 and 34.47 respectively, this refers to the petroleum exploitation existing in these places as well as the high values of CO_2, which, in turn, increase the index value in a comparison with the other sampling stations.

It is clear that, however, the highest values of the air pollution index with 69.30 and 41.63 have been recorded in the sampling stations of Burchesya and Seeba, respectively. For the former, this is due to the gaseous pollutants emitted from Abadan refinery offshore this site, in addition to there are a newer exploitation of petroleum within it; for the latter, this increase related to the gaseous emissions from dense works of petroleum exploitation on this site. The sampling station of Ras Abasha was the less polluted with an index value of 21.38, because of its marine nature and relatively distant from the influence of anthropocentric pollution sources.

CONCLUSIONS

In accordance to the spatial analysis of the geographical distribution of air pollutants, the present study concluded that the levels of air pollution in Basra province (the study area) were spatially varied, and that this spatial variation consequently correlated with the geographical distribution of gaseous emission sources existing in the study area. However, there is an irregular spatial pattern of air pollution concentration, although of some hotspots such as Burchesya and Seeba.

The authors recommend that mitigation to air pollution sources, particularly from stationary emission sources in the study area, is urgent action. Moreover, the installing of fixed stations designated to the air pollution monitoring is so necessary. Nevertheless, the accurate assessment of air pollution status in the study area needs more research and monitoring.

ACKNOWLEDGEMENTS

The authors acknowledge the following labs for the technical assistants: Environmental Analysis and Research Lab in Department of Geography, College of Arts, University of Basra; Gas Chromatography Lab in Department of Environmental Chemistry, Marine Science Center, University of Basra, and Air Pollution Lab in Department of Environmental and Pollution Engineering, Technical College, Southern Technical University.

REFERENCES

1. Admassu, M. and Wubeshet, M. (2006) Air Pollution: Lecture Notes for Environmental Health Science Students. University of Gondar Publications, Ethiopia, 5-6. http://www.cartercenter.org/resources/pdfs/health/ephti/library/lecture_notes/env_health_science_students/AirPollution.pdf

2. Matthias, A.D., Comrie, A.C. and Musil, S.A. (2006) Atmospheric Pollution. In: Peeper, I.L., Gerba, C.P. and Brusseau, M.L., Eds., Environmental and Pollution Science, 2nd Edition, Elsevier, San Diego, 377-394.

3. Vallero, D.A. (2008) Fundamentals of Air Pollution. 4th Edition, Elsevier Inc., London, 3.

4. Spellman, F.R. (1999) The Science of Environmental Pollution. Taylor & Francis Routledge, Pennsylvania, 245.

5. Harrison, R.M. (2001) Air Pollution: Sources, Concentrations and Measurements. In: Harrison, R.M., Ed., Pollution: Causes, Effects

and Control, 4th Edition, RSC, Cambridge, 169-192. http://dx.doi.org/10.1039/9781847551719-00169

6. Gittins, M.J. (1999) Air Pollution. In: Bassett, W.H., Ed., Clay's Handbook of Environmental Health, 18th Edition, E & FN Spon, London, 729-776. http://dx.doi.org/10.4324/9780203016312.ch42

7. Christoforou, C. (2004) Air Pollution. In: Stapleton, R.M., Ed., Pollution A to Z, Vol. 1, Macmillan Reference, New York, 30-38.

8. Al-Asadi, K.A.W. (1998) The Influence of Climatic Factors on the Major Industries in Basra and Their Reflections on the Environmental Pollution. Ph.D. Thesis, College of Arts, University of Basra, 197. (In Arabic)

9. Al-Mayahi, I.K. (2005) An Environmental Analysis on the Factors Affecting the Air Pollutants Quality at Basra Province. M.A. Thesis, College of Education, University of Basra, 240. (In Arabic)

10. Al-Imarah, F.J.M., Al-Mohameed, R.S.J. and Ibraheem, S.I. (2007) Extent of Atmospheric Pollution by Some Industrial Emissions Released from Petrochemicals and Gas Liquefier Industries in Khor Al-Zubair. Journal of Kerbala, Special Issue on the Annual Environment of Meeting of Babylon University, 1-6.

11. Al-Saad, H.T., Al-Imarah, F.J.M., Hassan, W.F., Jasim, A.H. and Hassan, I.F. (2010) Determination of Some Trace Elements in the Fallen Dust on Basra Governorate. Basrah Journal of Science, 28, 243-252. http://www.iasj.net/iasj?func=fulltext&aId=55085

12. Garabedian, S.A.K. (2010) Study of the Main Pollutants of Air Caused by Transportation in Basra. Proceedings of the Scientific Conference of Marine Science Center and the National Specialized Workshop on Oceanography, 23-24 December 2008, 125-136. (In Arabic)

13. Al-Hassen, Sh.I. (2011) Environmental Pollution in Basra City. PhD Thesis, College of Arts, University of Basra, Basra, 232. (In Arabic)

14. Qassim, M.H. (2011) A Geographic Analysis for Air Pollution Problem in Al-Zubayr City and Its Healthy Effects. M.A. Thesis, College of Arts, University of Basra, Basra, 180. (In Arabic)

15. Douabul, A.A.Z., Al-Maarofi, S.S., Al-Saad, H.T. and Al-Hassen, Sh.I. (2013) Gaseous Pollutants in Basra City, Iraq. Air, Soil and Water Research, 6, 15-21. http://dx.doi.org/10.4137/ASWR.S10835

16. Sultan, A.W.A., Al-Hassen, Sh.I., Ateeq, A.A. and Al-Saad, H.T. (2013) Ambient Air Quality in the Industrial Area of Khor Al-Zubayr, Southern Iraq. Journal of Petroleum Research & Studies, 4, 1-11. http://www.iasj.net/iasj?func=fulltext&aId=87876

17. Karmalla, H.A., Al-Hassen, Sh.I., Adam, R.S. and Qassim, M.H. (2013) A Cartographic Representation to Levels and Impacts of Carbon Monoxide Pollutant in Basra City, Southern Iraq. Journal of Thi-Qar Science, 4, 105-115. (In Arabic) http://www.iasj.net/iasj?func=fulltext&aId=83346

18. Abdullah, A.S. and Hussien, H.H. (2014) Estimation of Gaseous Pollutants emitted from Al-Najybia Power Station in Basra. Journal of Thi-Qar Science, 4, 68-71. http://www.iasj.net/iasj?func=fulltext&aId=92673

19. Al-Hassen, Sh.I., Al-Qarroni, E.H., Qassim, M.H., Al-Saad, H.T. and Alhello, A.A. (2015) An Experimental Study on the Determination of Air Pollutant Concentrations released from Selected Outdoor Gaseous Emission Sources in Basra City (Southern Iraq). JIARM, 3, 88-98. http://www.jiarm.com/FEB2015/paper19552.pdf

20. Wikipedia-Basra. https://en.wikipedia.org/wiki/Basra

21. ATSDR (Agency for Toxic Substances and Disease Registry), Toxic Substances Portal-Carbon Monoxide. http://www.atsdr.cdc.gov/toxfaqs/TF.asp?id=1163&tid=253

22. WHO (World Health Organization) (2010) WHO Guidelines for Indoor Air Quality: Selected Pollutants. WHO Regional Office for Europe, Bonn, 454. http://www.euro.who.int/__data/assets/pdf_file/0009/128169/e94535.pdf

23. ATSDR (Agency for Toxic Substances and Disease Registry), Toxic Substances Portal-Nitrogen Oxide. http://www.atsdr.cdc.gov/toxfaqs/TF.asp?id=396&tid=69

24. ATSDR (Agency for Toxic Substances and Disease Registry), Toxic Substances Portal-Sulfate Oxide. http://www.atsdr.cdc.gov/toxfaqs/TF.asp?id=252&tid=46

25. ATSDR (Agency for Toxic Substances and Disease Registry), Toxic Substances Portal-Hydrogen Sulfide. http://www.atsdr.cdc.gov/substances/toxsubstance.asp?toxid=67

26. ATSDR (Agency for Toxic Substances and Disease Registry), Toxic Substances Portal-Methane. http://www.atsdr.cdc.gov/HAC/landfill/html/ch4.html

27. ATSDR (Agency for Toxic Substances and Disease Registry), Toxic Substances Portal-Formaldehyde. http://www.atsdr.cdc.gov/mmg/mmg.asp?id=216&tid=39

28. WHO (World Health Organization) (2005) WHO Air Quality Guidelines for Particulate Matter, Ozone, Nitrogen Dioxide and Sulfur Dioxide: Global Update. Geneva, 20. http://whqlibdoc.who.int/hq/2006/who_sde_phe_oeh_06.02_eng.pdf

Chapter 9

SIGNIFICANCE OF PERSONAL EXPOSURE ASSESSMENT TO AIR POLLUTION IN THE URBAN AREAS OF EGYPT

Mahmoud M. M. Abdel-Salam

Department of Environmental Sciences, Faculty of Science, Alexandria University, Alexandria, Egypt

ABSTRACT

Air pollution, both indoor and outdoor, has been found to be related to serious adverse health effects. Accurate estimation of air pollution exposure has become very important to suggest proper air pollution control policies and to further assess the effectiveness of these policies. In many instances, personal exposures have been found to be greater than concentrations measured at fixed site monitoring stations. As people spend most of their time indoors particularly during harsh weather conditions, it is necessary to consider indoor air quality in exposure assessment studies. The current paper focuses on the importance of personal exposure assessment based on spatial and temporal activity patters both indoors and outdoors.

INTRODUCTION

A primary goal of air pollution control policy is to reduce, or ideally eliminate adverse effects on human health. It is well known that health risk assessment is most effective as a policy tool when it is linked to realistic exposure assessment, rather than to environmental concentrations. This is because exposure (defined in the context of this application as direct contact between an air pollutant and a human being [1]) is likely to be more closely related to health effects experienced by individuals. However, many study designs used in time series studies of health effects, and hence in assessing the health benefits of emission reductions, relied on air pollution concentrations measured at fixed measurement stations, usually at urban background sites, to define exposure of the study populations [2] - [4] . However, it is well established by extensive modeling and measurement studies that these concentrations cannot represent actual population exposures, because of the huge temporal and spatial variation in human activities [5] [6] . This creates significant difficulties in establishing the real health benefits of policies designed to meet ambient air quality standards which are based on these time series studies [7] .

The development of air pollution control policy requires accurate assessment of the impacts of current concentrations, and of the potential benefits of measures to reduce current emissions. In the case of human health, a key element of any such assessment is an estimate of the actual exposure of individuals to specific pollutants. While the measurement of air pollution is usually based on the use of fixed site monitors, the mobility of individuals means that their actual exposure will vary over the course of a day, as they move among locations such as home, street, office, car, etc. This interaction between the spatial and temporal variation in pollutant concentration and the time-activity patterns of individual people means that almost every individual will have a unique personal exposure to air pollution (i.e. unique concentration profile). For example, [8] found that the individual exposure to particulate matter of diameter less than 10 μm (PM_{10}) of a small group of traffic wardens differed by a factor of 5. The rapid changes in concentration as the individual moves from one location to another may have health significance. Also, cumulative exposure, or average exposure over certain fixed time intervals may be more important considerations.

There is a well-established association between ambient air pollution and adverse health outcomes in developed and developing countries [9] - [13] . The probability of adverse health outcomes will be greater in those individuals with higher exposure. In most urban cities, the highest outdoor concentrations are found at roadside sites, and thus the greatest health benefits may appear to arise from measures which reduce roadside concentrations. Indeed, there is evidence in many studies that the greater exposure to air pollutants of individuals living close to roads is associated with a greater prevalence of respiratory symptoms [14] - [16] . For example, different studies in the US and Europe have reported significant associations between daily mortality and PM_{10} [17] - [19] . It has been found that both respiratory and cardiovascular diseases are the major causes of death. Many studies have also shown positive associations between daily PM_{10} concentrations and daily hospital admissions for respiratory diseases (e.g. asthma, pneumonia, chronic obstructive pulmonary disease (COPD), etc.) [20] [21] . Other studies have reported increased hospital admissions for cardiovascular diseases (e.g. congestive heart failure, increase in coronary artery disease, etc.) associated with increased particle concentrations [22] .

However, evaluation of the significance of such health effects within the wider urban population depends critically on how roadside concentrations at specific sites actually relate to population-scale exposures. While ambient air quality data from individual urban background monitoring stations are typical of significant areas of cities, a major problem in the interpretation of data from roadside stations is that they are heavily influenced by the characteristics of local traffic flow. This means that it is problematic to use data from such sites in population-based estimates of impacts on personal exposure and health outcomes. Furthermore, people spend most of their time indoors particularly for groups such as elderly or those with pre-existing disease. Therefore, it is essential for estimates of population exposure that the relationships between indoor and ambient concentrations at both roadside and urban background sites are considered. Previous studies clearly show that even at roadside sites, indoor concentrations are substantially modified by the effect of indoor sources [23] - [25] .

Direct personal measurement and the use of diaries by participants can be used to estimate exposure and indeed to demonstrate closer links of health effects to exposure than to outdoor concentrations [26] . Methods of measuring individual pollutant exposure include personal (direct) method and microenvironmental (indirect) method. Personal monitoring involves attaching a monitor to an individual. The microenvironmental approach involves measuring pollutant concentrations in a number of defined environments in which an individual spends time, and combining these measurements with information on the person's time-activity pattern to estimate their exposure. Both personal and semi-portable background monitors may be used to estimate pollutant concentrations in the various environments. Personal sampling is the most reliable means of assessing true exposure and it is used to evaluate as accurately as possible an individual's exposure to air pollutants. Exposure measurements are particularly important in both workplace environments and in recent studies of air pollution exposures [27] - [36] . Information on the person's time-activity pattern can also be collected in order to interpret the person's exposure profile. Time-activity data (e.g. time spent commuting to and from work, mode of transport, smoking characteristics, etc.) may be collected by direct observation, by the use of self-administered daily time-activity diaries, or by questionnaires. However, when considering air pollution control policy over a nation, a region or a city, it is necessary to consider the exposure of the population as a whole. If the objective is to interpret population-based studies or to estimate health effects or health benefits across a population, then this direct approach is not possible, and the only practical approach to estimating exposure across the population, rather than individuals, is through appropriate computer models.

The nature of an individual's or a population's behavior, in terms of spatial and temporal activity patters, is a crucial determinant of their personal exposure to air pollutants. Because personal exposure assessment is concerned with the exposure to a single pollutant at the level of the individual, any pollutant monitoring exercise must incorporate the spatial and temporal variables associated with both individuals and pollutants. The exposure to any one pollutant will vary markedly depending on the location, characteristics of that location, activities undertaken, time of day or season, and the actual time spent in that location [37] . In many cases, personal exposures have been found to be greater than concentrations measured at fixed site monitoring stations [27] [38] - [40] .

While the development of personal exposure assessment methodologies has been a key element of research in the US over the past three decades, it has received little or no attention in most developing countries as Egypt. Personal exposure assessment is the only basis on which the health effects of air pollutants on individuals or populations can properly be assessed. The results of personal exposure assessment have often provided new thoughts into the importance of different sources of exposure, which are significant in terms of developing effective control strategies. For example, [41] estimated that, for benzene, 50% of the population exposure was due to direct smoking, 25% was due to personal activities such as passive smoking, car driving and the use of certain consumer products, and only 15% due to direct outdoor exposure. Thus, measures to reduce outdoor concentrations may have rather limited impact on population exposure, unless they are accompanied by measures to reduce the other sources of exposure. Also, US studies of personal exposure to CO have suggested that the most important element of exposure is that within vehicles; this may also be an important source of exposure to other pollutants, such as benzene and other VOCs [42] .

Exposure assessment can help us to determine what, where, and when pollutants come in contact with humans. The assessment of personal exposure is also of critical importance in terms of objectively assessing the relative benefits of different pollution control policies [43] . Unless the relative importance of exposure in different locations can be evaluated, it is impossible to properly assess the implications of policies to reduce exposure in individual locations. As Egypt has promulgated ambient air quality standards established by the Egyptian Environmental Law No. 4, 1994, and developed policies to meet these standards to protect public health, it is essential that the benefits of these policies for personal exposure can be properly assessed. The current procedures of comparing concentrations at fixed site monitoring stations with these standards, or estimating the number of people living in areas with particular pollutant concentrations are clearly inadequate for this purpose.

METHODOLOGY AND SIGNIFICANCE

A series of exposure assessment studies are to be conducted in Egypt to quantify and assess exposure of different groups of people including both normal and sensitive (e.g., children and elderly) groups to air pollutants. This will identify activities and microenvironments that contribute most to daily exposures of participants. Personal exposure will be measured in the different locations in which people spend time such as homes, cars, offices, schools, elderly care centers, etc. Personal monitors of particulate matter (PM_{10} and $PM_{2.5}$) and gaseous air pollutants known to cause adverse health effects (e.g. CO, NO_2, VOCs, etc.) will be used to continuously measure personal exposure by either direct attachment to participants or being placed in different microenvironments. Electronic diaries will be used to determine time-activity patterns and exposure profiles of all subjects. Conducting the studies in different areas and seasons will demonstrate the temporal and spatial variation of exposure levels of different groups. Outdoor air quality data will be collected to assess how personal exposures differ from outdoor concentrations and that the impacts of air pollution on an individual's health actually relate not to these outdoor concentrations but to their personal exposure.

It is highly expected to find a wide variation in personal exposure among different groups as well as individuals of the same groups as a result of different indoor and outdoor emission sources and activity patterns. Identifying groups of people with high personal exposure and their underlying causes is particularly important in Egypt where emission levels are high, but there are limited resources for environmental and health protection. This can identify a wider range of policy measures to significantly reduce exposure of such groups than direct emission control. As the science of personal exposure assessment has increasingly developed in North America and Western Europe, there is an urgent need to apply this science, with the associated measurement and modeling techniques, in other countries where the impacts of air pollution are much more serious. The aims of the current paper are to emphasize the significance of the followings:

- Introduce and establish a personal exposure assessment strategy using direct measurement (i.e. personal sampling technique) and indirect methods (i.e. microenvironmental method) as the only basis on which the health effects of air pollutants on individuals or populations can reliably be assessed;

- Properly evaluate personal exposure in different locations (both indoors and outdoors) to air pollutants and thus to accurately assess current policies which are used to reduce emissions in different sectors in terms of their impacts on human health, i.e. objectively assessing the relative benefits of different pollution control policies;

- Select and develop the most effective air pollution control strategies that can significantly reduce exposure to air pollutants;

- Construct a database of personal exposure measurements that is useful in estimating exposure across the population through appropriate computer models;

- Set proper air quality standards, both indoor and outdoor, according to the accurate and reliable estimates of personal exposure as well as the related health impacts;

- Assist in advising national, regional and local government in developing stringent policies and actions that will reduce urban air pollution.

In the light of all of the above, this will help to introduce and establish the process of personal exposure assessment to air pollution in Egypt as an effective tool and the only basis on which the health effects of air pollutants on individuals and population can properly be assessed. Also, this will enhance our knowledge about the pattern of human exposures in Egypt, particularly in urban areas where high concentrations of air pollutants and related health impacts always exist. This will help to save and protect human health through the development of effective air pollution control strategies and also contribute to raise awareness of the public regarding exposure assessment through their potential participation as volunteers in related procedures.

REFERENCES

1. Zartarian, V.G., Ott, W.R. and Duan, N. (1997) A Quantitative Definition of Exposure and Related Concepts. Journal of Exposure Analysis and Environmental Epidemiology, 7, 411-438.

2. Katsouyanni, K., Schwartz, J., Spix, C., Touloumi G., Zxirou, D., Zanobetti, A., Wojtyniak, B., Vonk, J.M., Tobias, A., Ponka, A., Medina, S., Bacharova, L. and Anderson, H.R. (1996) Short Term Effects of Air Pollution on Health: A European Approach Using Epidemiologic Time Series Data: The APHEA Protocol. Journal of Epidemiology and Community Health, 50, S12-S18. http://dx.doi.org/10.1136/jech.50.Suppl_1.S12

3. Schwartz, J., Spix, C., Touloumi G., Bacharova, L., Barumamdzadeh, T., le Tetre, A., Tierkarski, T., Ponce de Leon, A., Ponka, A., Rossi, G., Saez, M. and Schouten, J.P. (1996) Methodological Issues in Studies of Air Pollution and Daily Counts of Deaths or Hospital Admissions. Journal of Epidemiology and Community Health, 50, S3-S11.http://dx.doi.org/10.1136/jech.50.Suppl_1.S3

4. Atkinson, R.W., Anderson, H.R., Strachan, D.P., Bland, J.M., Bremner, S.A. and de Leon, A.P. (1999) Short Term Associations between Emergency Hospital Admissions for Respiratory and Cardiovascular Disease and Outdoor Pollution in London. Archives of Environmental Health, 54, 398-411. http://dx.doi.org/10.1080/00039899909603371

5. Wallace, L.A. (1995) Human Exposure to Environmental Pollutants: A Decade of Experience. Clinical and Experimental Allergy, 25, 4-9. http://dx.doi.org/10.1111/j.1365-2222.1995.tb00996.x

6. Violante, F.S., Barbieri, A., Curti, S., Sanguinetti, G., Graziosi, F. and Mattioli, S. (2006) Urban Atmospheric Pollution: Personal Exposure versus Fixed Monitoring Station Measurements. Chemosphere, 64, 1722-1729.http://dx.doi.org/10.1016/j.chemosphere.2006.01.011

7. Bahadori, T., Suh, H. and Koutrakis, P. (1999) Issues in Human Particulate Exposure Assessment: Relationship between Outdoor, Indoor and Personal Exposures. Human and Ecological Risk Assessment, 5, 459-470.http://dx.doi.org/10.1080/10807039.1999.10518871

8. Watt, M., Godden, D., Cherrie, J. and Seaton, A. (1995) Individual Exposure to Particulate Air Pollution and Its Relevance to Thresholds for Health Effects. Occupational and Environmental Medicine, 52, 790-792. http://dx.doi.org/10.1136/oem.52.12.790

9. Dockery, D.W., Pope, C.A., Xu, X., Spengler, J.D., Ware, J.H., Fay, M.E., Ferris, B.G. and Speizer, F.E. (1993) An Association between Air Pollution and Mortality in Six U.S. Cities. New England Journal of Medicine, 329, 1753-1759.http://dx.doi.org/10.1056/NEJM199312093292401

10. Schwartz, J. (1994) Air Pollution and Daily Mortality: A Review and Meta Analysis. Environmental Research, 64, 36- 52. http://dx.doi.org/10.1006/enrs.1994.1005

11. WHO (2001) World Health Report. World Health Organization, Geneva.

12. WHO (2002) World Health Report: Reducing Risks, Promoting Healthy Life. World Health Organization, Geneva.

13. Pope, C.A. and Dockery, D.W. (2006) Health Effects of Fine Particulate Air Pollution: Lines that Connect. Journal of the Air & Waste Management Association, 56, 709-742.http://dx.doi.org/10.1080/10473289.2006.104 64485

14. Van Vliet, P., Knape, M., de Hartog, J., Janssen, N.A.H., Harssema, H. and Brunekreef, B. (1997) Air Pollution from Road Traffic and Chronic Respiratory Symptoms in Children Living near Motorways. Environmental Research, 74, 122-132.http://dx.doi.org/10.1006/enrs.1997.3757

15. Ciccone, G., Forastiere, F., Agabiti, N., Biggeri, A., Bisanti, L., Chellini, E., Corbo, G., Dell'Orco, V., Dalmasso, P., Volante, T.F., Galassi, C., Piffer, S., Renzoni, E., Rusconi, F., Sestini, P. and Viegi, G. (1998) Road Traffic and Adverse Respiratory Effects in Children. Occupational and Environmental Medicine, 55, 771-778.http://dx.doi.org/10.1136/oem.55.11.771

16. Kramer, U., Koch, T., Ranft, U., Ring, J. and Behrendt, H. (2000) Traffic-Related Air Pollution Is Associated with Atopy in Children Living in Urban Areas. Epidemiology, 11, 64-70. http://dx.doi.org/10.1097/00001648-200001000-00014

17. Schwartz, J. (1993) Air Pollution and Daily Mortality in Birmingham, Alabama. American Journal of Epidemiology, 137, 1136-1147.

18. Verhoeff, A.P., Hoek, G., Schwartz, J. and van Wijuen, J.H. (1996) Air Pollution and Daily Mortality in Amsterdam, the Netherlands. Epidemiology, 7, 225-230.http://dx.doi.org/10.1097/00001648-199605000-00002

19. Janssen, N.A.H., Fischer, P., Marra, M., Ameling, C. and Cassee, F.R. (2013) Short-Term Effects of $PM_{2.5}$, PM_{10} and $PM_{2.5-10}$ on Daily Mortality in the Netherlands. Science of the Total Environment, 464, 20-26. http://dx.doi.org/10.1016/j.scitotenv.2013.05.062

20. Schwartz, J., Slater, D., Larson, T.V., Pierson, W.E. and Koenig, J.Q. (1993) Particulate Air Pollution and Hospital Emergency Room Visits for Asthma in Seattle. American Review of Respiratory Disease, 147, 826-831. http://dx.doi.org/10.1164/ajrccm/147.4.826

21. Dockery, D.W. and Pope, C.A. (1994) Acute Respiratory Effects of Particulate Air Pollution. Annual Review of Public Health, 15, 107-132. http://dx.doi.org/10.1146/annurev.pu.15.050194.000543

22. Schwartz, J. and Morris, R. (1995) Air Pollution and Hospital Admissions for Cardiovascular Disease in Detroit, Michigan. American Journal of Epidemiology, 142, 23-35.

23. Jones, N.C., Thornton, C.A., Mark, D. and Harrison, R.M. (2000) Indoor/Outdoor Relationships of Particulate Matter in Domestic Homes with Roadside, Urban and Rural Locations. Atmospheric Environment, 34, 2603-2612. http://dx.doi.org/10.1016/S1352-2310(99)00489-6

24. Abdel-Salam, M.M. (2012) Indoor Particulate Matter in Different Residential Areas of Alexandria City, Egypt. Indoor & Built Environment, 21, 857-862.http://dx.doi.org/10.1177/1420326X11422262

25. Abdel-Salam, M.M. (2013) Indoor Particulate Matter in Urban Residences of Alexandria, Egypt. Journal of the Air & Waste Management

Association, 63, 956-962.http://dx.doi.org/10.1080/10962247.2013.801374

26. Seaton, A., Soutar, A., Crawford, V., Elton, R., McNerlan, S., Cherrie, J., Watt, M., Agius, R. and Stout, R. (1999) Particulate Air Pollution and the Blood. Thorax, 54, 1027-1032.http://dx.doi.org/10.1136/thx.54.11.1027

27. Spengler, J.D., Treitman, R.D., Tosteson, T.D., Mage, D.T. and Soczek, M.L. (1985) Personal Exposures to Respirable Particulates and Implications for Air Pollution Epidemiology. Environmental Science and Technology, 19, 700-707.http://dx.doi.org/10.1021/es00138a008

28. Wallace, L. (1993) A Decade of Studies of Human Exposure: What Have We Learned. Risk Analysis, 13, 135-139. http://dx.doi.org/10.1111/j.1539-6924.1993.tb01059.x

29. Chan, L.Y., Kwok, W.S. and Chan, C.Y. (2000) Human Exposure to Respirable Suspended Particulate and Airborne Lead in Different Roadside Microenvironments. Chemosphere, 41, 93-99. http://dx.doi.org/10.1016/S0045-6535(99)00394-X

30. Janssen, N.A.H., van Vliet, P.H.N., Aarts, F., Harssema, H. and Brunekreef, B. (2001) Assessment of Exposure to Traffic Related Air Pollution of Children Attending Schools near Motorways. Atmospheric Environment, 35, 3875- 3884. http://dx.doi.org/10.1016/S1352-2310(01)00144-3

31. Rojas-Bracho, L., Suh, H.H., Oyola, B. and Koutrakis, P. (2002) Measurements of Children's Exposures to Particles and Nitrogen Dioxide in Santiago, Chile. Science of the Total Environment, 287, 249-264. http://dx.doi.org/10.1016/S0048-9697(01)00987-1

32. Mukherjee, A.K., Bhattacharya, S.K., Ahmed, S., Roy, S.K., Roychowdhury, A. and Sen, S. (2003) Exposure of Drivers and Conductors to Noise, Heat, Dust and Volatile Organic Compounds in the State Transport Special Buses of Kolkata City. Transportation Research, D8, 11-19. http://dx.doi.org/10.1016/S1361-9209(02)00015-9

33. Gulliver, J. and Briggs, D.J. (2004) Personal Exposure to Particulate Air Pollution in Transport Microenvironments. Atmospheric Environment, 38, 1-8.http://dx.doi.org/10.1016/j.atmosenv.2003.09.036

34. Kaur, S., Nieuwenhuijsen, M.J. and Colvile, R.N. (2005) Pedestrian Exposure to Air Pollution along a Major Road in Central London, UK. Atmospheric Environment, 39, 7307-7320. http://dx.doi.org/10.1016/j.atmosenv.2005.09.008

35. Dons, E., Panis, L., Poppel, M., Theunis, J., Willems, H., Torfs, R. and Wets, G. (2011) Impact of Time-Activity Patterns on Personal Exposure

to Black Carbon. Atmospheric Environment, 45, 3594-3602. http://dx.doi.org/10.1016/j.atmosenv.2011.03.064

36. Steinle, S., Reis, S. and Sabel, C. (2013) Quantifying Human Exposure to Air Pollution—Moving from Static Monitoring to Spatio-Temporally Resolved Personal Exposure Assessment. Science of the Total Environment, 443, 184-193.http://dx.doi.org/10.1016/j.scitotenv.2012.10.098

37. Loth, K. and Ashmore, M. (1994) Assessment of Personal Exposure to Air Pollution. Clean Air, 24, 114-122.

38. Letz, R., Ryan, P.B. and Spengler, J.D. (1984) Estimated Distribution of Personal Exposure to Respirable Particles. Environmental Monitoring Assessment, 4, 351-359.http://dx.doi.org/10.1007/BF00394173

39. Ashmore, M. and Dimitroulopoulou, C. (2009) Personal Exposure of Children to Air Pollution. Atmospheric Environment, 43, 128-141.http://dx.doi.org/10.1016/j.atmosenv.2008.09.024

40. An, X., Hou, Q., Li, N. and Zhai, S. (2013) Assessment of Human Exposure Level to PM_{10} in China. Atmospheric Environment, 70, 376-386.http://dx.doi.org/10.1016/j.atmosenv.2013.01.017

41. Wallace, L.A. (1989) Major Sources of Benzene Exposure. Environmental Health Perspectives, 82, 165-169. http://dx.doi.org/10.1289/ehp.8982165

42. Chan, C., Spengler, J., Ozkaynak, H. and Lefkopoulou, M. (1991) Commuter Exposures to VOCs in Boston, Massachusetts. Journal of the Air & Waste Management Association, 41, 1594-1600. http://dx.doi.org/10.1080/10473289.1991.10466955

43. Steinemann, A. (2004) Human Exposure, Health Hazards, and Environmental Regulations. Environmental Impact Assessment Review, 24, 695-710.http://dx.doi.org/10.1016/j.eiar.2004.06.002

Chapter 10

TRACE METALS CONCENTRATIONS AT THE ATMOSPHERE PARTICULATE MATTERS IN THE SOUTHEAST ASIAN MEGA CITY (DHAKA, BANGLADESH)

Md. Faridul Islam[1], Syada Sanjida Majumder[1], Abdullah Al Mamun[1,2], Md. Badiuzzaman Khan[3,4], Mohammad Arifur Rahman[1], Abdus Salam[1]

[1]Department of Chemistry, University of Dhaka, Dhaka, Bangladesh

[2]Department of Chemistry, University of Louisville, Louisville, USA

[3]Department of Environment Sciences Informatics and Statistics, Cà Foscari University of Venice, Venice, Italy

[4]Department of Environment Science, Bangladesh Agricultural University, Mymensingh, Bangladesh

ABSTRACT

Atmospheric particulate matters were collected on quartz fibre filters for 24 hours with a low volume sampler from January 2014 to March 2014 at the Southeast Asian mega city (Dhaka, Bangladesh). Particulate matters samples were analysed for eleven trace metals with inductively coupled plasma mass spectrometer (ICP-MS) at Cà Foscari University of Venice, Italy. Trace metals were extracted from filters with digestion method using a mixture of HNO_3 and H_2O_2. The average concentration of the determined trace metals of As, Cd, Ni, Cu, Pb, Cr, Fe, Mn, Zn, Sband Se were 3.06, 6.28, 3.77, 11.98, 305.6, 9.2, 2057.0, 42.2, 303.3, 5.47 and 2.43 ng·m^{-3}, respectively. Arsenic concentration is much lower in the atmosphere of Dhaka, though Bangladesh has severe arsenic problem in the ground water. Lead and cadmium concentrations showed decreasing trend in Dhaka compared than previous measurements—but still they have very high levels compared than Europe and USA. There is very limited information for Mn, Sb and Se concentrations in Dhaka air. Correlation studies showed that several trace metals had potential joint sources of origin, e.g., manganese is highly correlated with iron ($r^2 = 0.97$) and nickel ($r^2 = 0.84$), copper ($r^2 = 0.86$); lead with arsenic ($r^2 = 0.79$) and antimony ($r^2 = 0.78$). Enrichment factors analysis was also done with the data base for the respective metals in earth crust and coal fly ash. As and Cu both have combined sources, whereas Cd, Pb and Zn were from coal fly ash.Trace metals concentrations in Dhaka city air were much higher than Europe and USA but comparable or slightly lower than other south Asian countries. This is the first extensive study for the eleven trace metals with ICP-MS in Dhaka, Bangladesh.

INTRODUCTION

Atmospheric particulate matter (PM) has significant impact on human health, climate change, visibility reduction, agriculture and atmospheric chemistry. Aerosol particles may include a range of chemical species, ranging from metals to organic and inorganic compounds [1] [2]. Among the inorganic compounds, most important ones are the trace metals, which are emitted by various natural and anthropogenic sources such as crustal materials, road dust, construction activities, motor vehicles, coal and oil combustion, incineration and other industrial activities [3] - [7]. Health impacts associated with particulate matters are linked to respiratory, cardiovascular problems, premature mortality, lung cancer, heart diseases and also damage to other organs [8] - [11]. Several studies have indicated that different transition metals may act as possible mediators of particle induced injury and inflammation [12] [13]. The attention has often been focused on transition metals such as iron, nickel, chromium, copper, and zinc, based on their ability to generate reactive oxygen species (ROS) in biological tissues [14]. Using single-component regression analysis, Gurgueira et al. [15] described that the content of Fe, Mn, Cu, and Zn was strongly associated with the oxidative stress generated in the lung, whereas Fe, Al, Si, and Ti were associated with the effects observed in the heart. The spatial and temporal variation in the risk of particulate matters is partially explained by chemical composition [16].

Atmospheric pollution is a serious public health problem in the developing countries especially Dhaka, Bangladesh due to the unplanned rapid growth of the city [17]. Dhaka, with about 17 million people and 8% increase of population per year [18], is exposed to the high levels of trace metal pollutions from a variety of sources [17]. Due to the presence of high-level toxic elements, Dhaka has been considered as one of the most polluted cities in the world [18]. Air pollution in Bangladesh is mainly caused by traffics, brick kilns, industries, biomass burning, construction activity, soil dust, and also long range transport, etc. [19]. Trace metals levels have the high exceedances of WHO guideline values in South Asian countries (e.g., India and Pakistan) [20]. The present environmental condition especially atmospheric pollution in Bangladesh is not at all equilibrium. Severe air, water and noise pollution are threatening human health, ecosystems and economic growth of Bangladesh. Several thousand people were died each year due to air pollution problem in Bangladesh. Thus, the monitoring of trace metals in atmospheric particulate matters in Dhaka city has become an essential part of environmental planning and control programmes in Bangladesh.

With this objective, we aimed for assessing particulate matters with respect to eleven trace metals (As, Cd, Ni, Cu, Pb, Cr, Fe, Mn, Zn, Sb and Se) between January and February 2014 in University of Dhaka, Bangladesh. Results from this work can provide useful information on trace metal composition of particulate matters. The results can also be used to support further studies on the impacts of traffic and industries generated pollutants on human health in the highly populated Dhaka City, Bangladesh.

EXPERIMENTAL METHOD

Description of the Sampling Location

Bangladesh is situated in the eastern part of south Asia. It is surrounded by India on the west, the north and the northeast, Myanmar on the southeast, and the Bay of Bengal on the south (Figure 1). Dhaka (Latitude: 23.72839° North, Longitude: 90.39819° East, Elevation: 34.0 Meters), the capital of Bangladesh, is the centre of commerce and industries for the country. Dhaka is growing rapidly and faces all the problems associated with mega-city. Dhaka is situated on flat land surrounded by rivers. The exact sampling location is situated on the roof of Mukarram Hussain Khundkur Science Building, Department of Chemistry, University of Dhaka, Bangladesh.

(a)

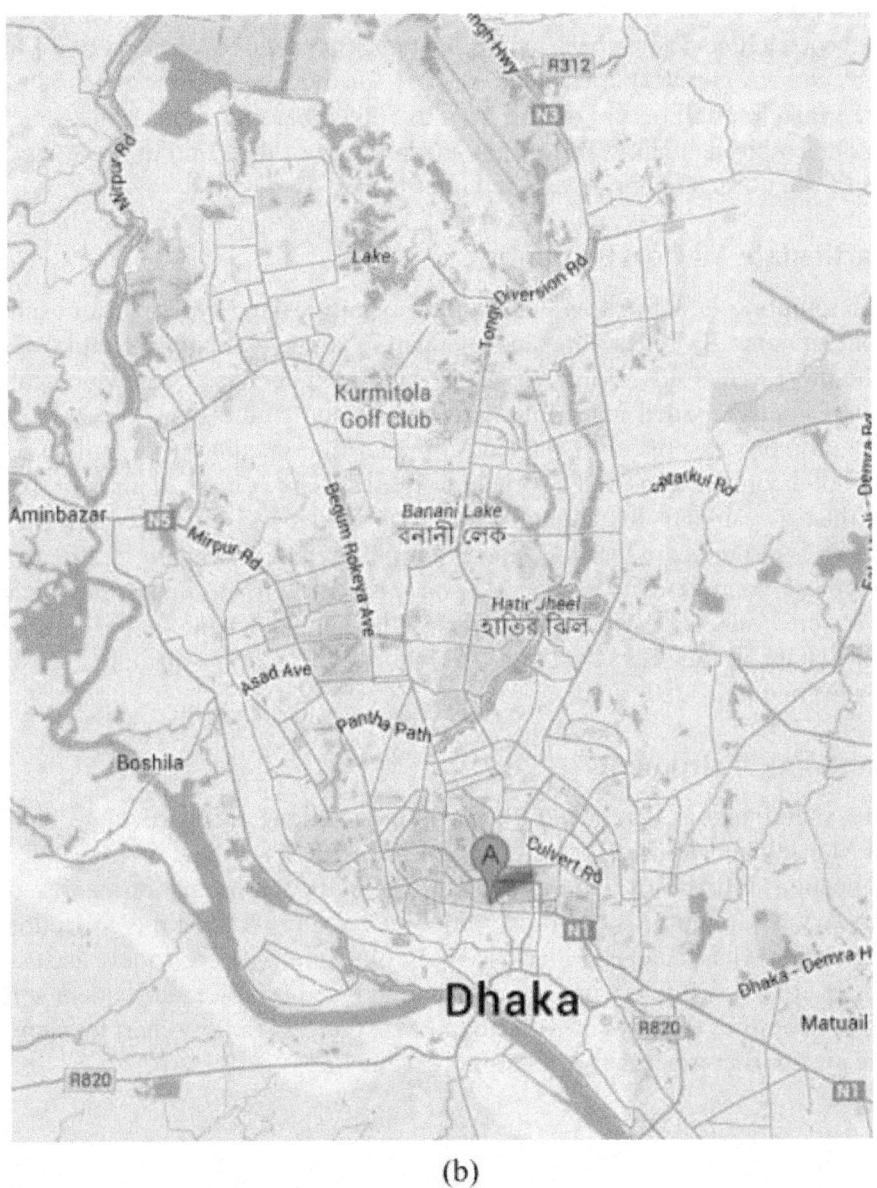

(b)

Figure 1. (a) Dhaka city map; (b) Map of Bangladesh.

Meteorology of the Sampling Location

The climate in Bangladesh is characterized by high temperature and high humidity most of the year with a distinct seasonal variation of precipitation.

The year can be divided into four seasons, pre-monsoon (March-May), monsoon (June-September), post monsoon (October-November) and winter (December-February) in Bangladesh [21]. On average, approximately 80% of the yearly rainfall occurs during May to September monsoon. Wind direction in Dhaka city is mainly from west and south-west direction at pre-monsoon, and from north and north-west at winter [22].

Particulate Matters Sampling

Air sampling is defined as determining quantities and types of atmospheric contaminants by measuring and evaluating a representative sample of air. The most numerous environmental hazards are chemical, ones which can be conveniently divided into a) the particulates and b) the gases or vapours. In filtration process, the air is passed through a filter medium (normally a paper for solid contaminants and a sorbent for gases). The volume of air is measured against the amount of contaminant captured. This gives the concentration, which is expressed either as Nano gram per meter cube ($ng \cdot m^{-3}$). Figure 2 is a typical example of the unloaded and loaded filters in Dhaka, Bangladesh. We were collected particulate matters with Gelman, Membrane Filters, Type TISSUQUARTZ, TISSUQUARTZ2500QAT-UP, 47mm diameter for both loaded and unloaded filters.

Sampling Instruments

The sampling was conducted by filtration technique and particulate sampler SPM machine was used for this sampling. Quartz fibre filters were used for air collection. Filtration sampling, which is actually a combination of filtration/impaction sampling, is the most widely used approach for the collection of atmospheric particulates. Filter-based sampling methods are widely used since filters are relatively low in cost, easily stored, and used for subsequent simple and/or complex analyses of collected SPM. Figure 3 is a flow chart for showing the air filtering process on the sampling filters.

(a)

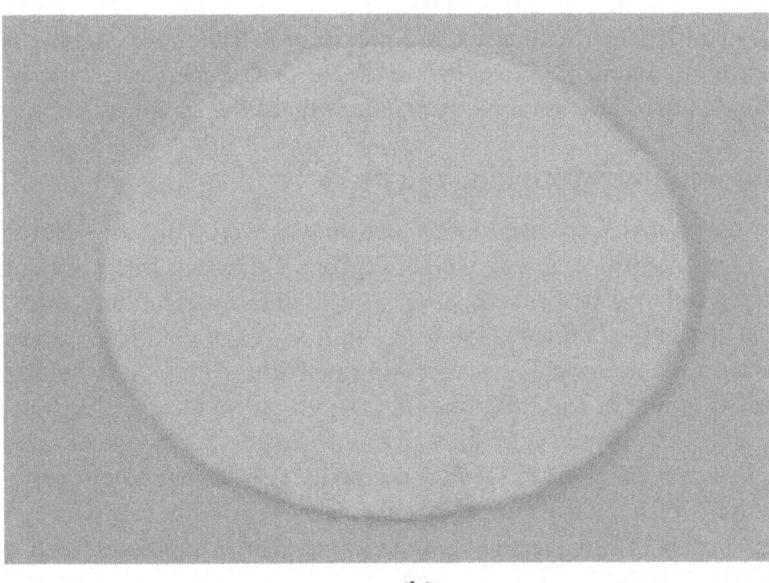

(b)

Figure 2. Loaded and unloaded filters at MHK Bhavan, Department of Chemistry, Dhaka, Bangladesh from January 2014 to March 2014. (a) 24-hour loaded filter; (b) Unloaded (blank) filter.

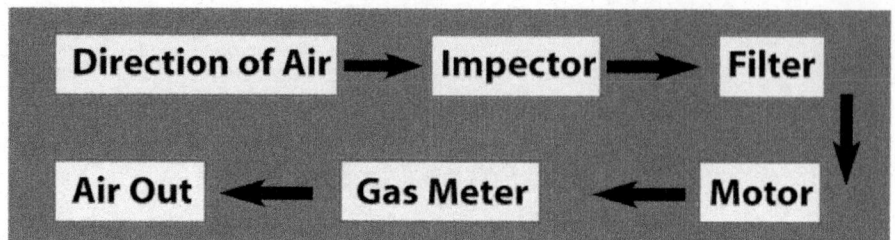

Figure 3. Flow chart of the air filtering system to collect particulate matter.

Description of Filtering of the Sample

Aerosol samples of SPM were collected daily in Dhaka from January to March in the winter and spring of 2014, using a low volume homemade sampler (Vienna, Austria). All samples were collected on the roof of the Mukarram Hussain Kkundkur Science Building, Department of Chemistry, Dhaka University using Quartz filter paper. The duration time of sampling was about 24-hour. The filters before and after sampling were weighed using an analytical balance with a reading precision 0.10mg after stabilizing under constant temperature (20°C ± 1°C). All the procedures were strictly quality controlled to avoid the possible contamination of the samples.

Conditioning and Weighing of Filters

All the filters that were used to collect sample were kept in a desiccator for 2 hours for conditioning. The weight of the conditioned filters was measured by a microbalance before exposure. The loaded quartz filters were kept in freeze at 4°C until chemical analysis to limit losses of volatile components. To reduce the organic species background level from filters, quartz filters were pre-treated at 800°C for 200 min and then placed in clean polyethylene Petri dishes. The Petri dishes were kept in the desiccator for two hours and finally stored in a freezer until field measurement. The loaded filters were sent to the Analytical Chemistry Laboratory for trace metals analysis with ICP-MS at the Department of Environment Sciences Informatics and Statistics, CàFoscari University of Venice, Italy.

Microwave Digestion and Elements Analysis

The half of the filter was digested by microwave digestion system (MARSX CEM Corporation USA) in a Teflon vessel with 8 ml 69% HNO_3 and 2 ml H_2O_2. The sample mixture was digested according to the CEN. EN 14902:2005 standards [23]. The digested solution was diluted to 50 ml with ultrapure water and stored at 4°C in PTFE bottles. Elemental analysis of As, Cd, Ni, Mn, Cu, Fe, Pb, Zn, Cr, Sb and Se were performed using inductively coupled plasma mass spectrometer (ICP-MS), model: 7700 Agilent, USA. Filter blanks and field blanks (filter kept inside the sampler but not used for air filtering) were prepared and analyzed together with the samples, following the same procedures and the values obtained were routinely subtracted from those of the samples. The limit of detection (LODs) was calculated for each ion and element as the three times the standard deviation of the blank values for each analyzed period, the limit of quantification (LOQ) is the lowest point of the calibration curve. The standards for the calibration curve were obtained by dilution from ICP multi element Standard Solution IV Merk and single element standard of VWR. The quality and accuracy of quantitative analysis were routinely checked analyzing the NIST SRM 1648 standard for air particulate. The recoveries of ions and elements were in the range of 80% - 110%.

RESULT AND DISCUSSION

All the collected samples from Dhakawere sent for analyses at the Air Pollution Laboratory, Regional Environmental Protection Agency (ARPAV), Venice, Italy. An inductively coupled plasma mass spectrometer (ICP-MS) was used to determine the concentrations of the trace metals (As, Cd, Cu, Fe, Pb, Zn, Cr, Se, Sb, Ni, Mn). The concentrations of these trace metals with average values have given in Table 1.

Arsenic

Arsenic (As) and its compounds are ubiquitous in nature and exhibit both metallic and nonmetallic properties. It is a major ground water pollutant in many parts of Bangladesh. Millions of people are affected by As-related diseases in Bangladesh due to contaminated drinking water. It is therefore very important to check the As concentrations in the air of Bangladesh. The total average As concentration found at Dhaka was 3.06 ng·m^{-3} varying from 1.50 to 4.20 ng·m^{-3} (Table 1), which was about compared than the previously reported concentration (6.3 ng·m^{-3}) in Dhaka by Salam et al. 2008. The allowable range for Arsenic in the United States was from 1.0 to 3.0 ng·m^{-3} for remote areas and 20 - 30 ng·m^{-3} for urban areas [24]. In England, the mean concentration was 5.4 ng·m^{-3} with a declining trend over the period 1957-1974 [25]. Arsenic in air is present mainly in particulate forms as inorganic arsenic. It is assumed that methylated arsenic is a minor component in the air of suburban, urban and industrial areas, and that the major inorganic portion is a variable mixture of the trivalent and pentavalent forms, the latter being predominant. Particulate arsenic compounds may be inhaled, deposited in the respiratory tract and absorbed into the blood. Arsenic is mostly coming from geogenic sources through it has some other sources.

Table 1. Concentration of atmospheric trace metals in the particulate matters in Dhaka, Bangladesh for the period January 2014 to March 2014. All units are in $ng \cdot m^{-3}$.

Elements	As	Cd	Ni	Pb	Cr	Cu	Fe	Mn	Zn	Sb	Se
	3.50	12.00	3.90	399.50	7.20	14.30	1665.50	38.70	497.70	6.10	2.90
	2.40	1.80	4.70	88.80	11.00	14.60	3207.70	61.70	257.90	2.10	1.60
	4.20	3.30	3.10	355.20	8.00	12.00	1727.70	34.70	304.40	9.20	3.20
	3.20	2.50	4.90	228.70	7.90	10.50	2071.30	38.50	88.10	6.10	2.50
	3.40	12.90	5.50	507.20	13.50	18.90	3022.10	65.10	474.60	5.10	3.40
	1.50	5.70	2.10	127.80	4.40	7.00	1242.60	22.20	99.90	1.50	1.20
	2.40	7.80	3.20	198.40	14.20	9.10	1920.00	40.10	481.90	3.00	2.60
	3.90	4.30	2.80	539.00	6.40	9.50	1598.70	36.50	221.80	10.70	2.10
Average	3.06	6.29	3.78	305.6	9.08	12.0	2057.0	42.2	303.3	5.5	2.44
Maximum	4.20	12.90	5.50	539.0	14.20	18.9	3207.7	65.1	497.7	10.7	3.40
Minimum	1.50	1.80	2.10	88.8	4.40	7.00	1242.6	22.2	88.1	1.50	1.20

Arsenic may also introduce in the air of Dhaka city through tobacco smoke and using pesticides containing Arsenic compound. It is estimated that the arsenic content of mainstream cigarette smoke is in the range 40 - 120 ng per cigarette. If consumption is 20 cigarettes per day, the daily intake from this source would amount to 0.8 - 2.4 µg. Moreover around the Dhaka city there are hundreds of Brick industries which also burning arsenic compound containing coal. Unexpectedly high As concentrations (325 $ng \cdot m^{-3}$) were also observed at all four sites in Bangkok, Thailand [26].

Cadmium

Average Cd concentration was found 6.28 $ng \cdot m^{-3}$ varying from 1.8 to 12.9 $ng \cdot m^{-3}$ (Table 1). The guideline value for Cadmium by WHO is 5.0 $ng \cdot m^{-3}$ which is slightly lower than the average value and may be comparable with our values, but three to fourfold lower than the values reported for other Southeast Asian sites [27] ranging from 0.2 to 34.9 $ng \cdot m^{-3}$. The total average of Cd concentration in Dhaka city was 13 $ng \cdot m^{-3}$ reported by Salam et al. [28], which is about two times higher than the current value. The sludge processing is one of the important sources for Cd. This lower value is indicating recent improvement of the sludge processing system in Dhaka city, Bangladesh.

Cadmium in ambient air represents the majority of total airborne cadmium. Inputs from different categories of cadmium may affect human Cdintake and human health, but the levels and the transfer mechanisms to humans are substantially different among them. Whereas cadmium from occupational environments and cadmium from cigarette smoke are transferred directly to humans, cadmium in ambient air is generally deposited onto waters or soils, then eventually transferred to plants and animals, and finally enter the human body through the food chain. Cadmium has high toxic effects. It affects kidneys and responsible for accumulating stones in kidneys.

Nickel

Nickel (Ni) is the 24th most abundant element in the Earth's crust, comprising about 3% of the composition of the earth. Nickel and nickel compounds have many industrial and commercial uses. Most of the Ni is used for the production of stainless steel and other nickel alloys with high corrosion and temperature resistance. Nickel metal and its alloys are used widely in the metallurgical, chemical and food processing industries, especially as catalysts and pigments. The sources of nickel are the paints and varnishes, dye, building material, electrical equipment and telephone cable, etc. The concentration of nickel is higher than any other location may causes of construction materials, paint and varnishes. Nickel is also uses for the galvanizing of metal plate. This may also be the cause of higher amount of nickel. In this study the average Nickel concentration was 3.77 $ng \cdot m^{-3}$varying from 2.1 to 5.5 $ng \cdot m^{-3}$ (Table 1), which is similar to the reported Nickel concentration in many parts of the world. In a remote area of Canadian Arctic, the Ni levels of 0.38 - 0.62 $ng \cdot m^{-3}$were recorded (Hoff and Barrie 1986), as compared to 124.0 $ng \cdot m^{-3}$ in the vicinity of a nickel smelter [29]. In northern Norway, a level of about 1.0 $ng \cdot m^{-3}$ was recorded in an unpolluted area as compared to about 5.0n gm^{-3} some 5 km distant from a nickel smelter (average values 1990-1991). The highest recorded value was 64 $ng \cdot m^{-3}$ [30] [31]. Concentrations of 18 - 42 $ng \cdot m^{-3}$were recorded in eight United States cities [32]. These values correspond to the average value of 37 $ng \cdot m^{-3}$ for 30 United States Urban Air National Surveillance Network Stations for the period 1957-1968. This average decreased from 47 $ng \cdot m^{-3}$ for 1957-1960 to 26 $ng \cdot m^{-3}$ for 1965-1968. The mean (arithmetic) value for 1970-1974 was 13 $ng \cdot m^{-3}$ [33]. Ranges of 10 - 50 $ng \cdot m^{-3}$ and 9 - 60 $ng \cdot m^{-3}$ have been reported in European cities. Higher values (110 - 180 $ng \cdot m^{-3}$) have been reported from heavily industrialized areas [34]. At Islamabad, Pakistan nickel concentration in TSP was 9.0 $ng \cdot m^{-3}$ and at Beijing, China nickel concentration in TSP was 51.0 $ng \cdot m^{-3}$ [35]. The nickel concentration in TSP at Yamaguchi, Japan was 16.0 $ng \cdot m^{-3}$ [36]. The average concentration of nickel observed in the air at Dhaka city is higher than the other Asian cities.

Copper

The main environmental issues for the primary copper production are the potential emission to air of dust and metals/metal compounds and of sulphur dioxide from roasting and smelting sulphide concentrates or using sulphur-containing fuels or other materials. The average Cu concentration was 11.98 $ng \cdot m^{-3}$ varying from 7.0 to 18.9 $ng \cdot m^{-3}$ (Table 1). Previously, Salam et al. [28] reported that the average concentration of copper at Dhaka University 31.0 $ng \cdot m^{-3}$, which is about two and half times higher than the current finding. The average copper concentration is showing decreasing tendency in Dhaka, the value of on 2003 was 54 $ng \cdot m^{-3}$ [21], which is twofold higher than the 2008 and fivefold higher than the current finding. However, it is a clear indication of decreasing Cu concentration in the air of Dhaka city. The high concentration of Cu (65 $ng \cdot m^{-3}$) was also observed at four sites in Bangkok, Thailand [26].

Lead

The main sources of lead are combustion of fossil fuels, paints and varnish, production of batteries, lead containing waste water. The average Pb concentration found in this current study 305.75 $ng \cdot m^{-3}$ varying from 88.6 to 539 $ng \cdot m^{-3}$ (Table 1). Lead compounds accumulate with PM in the atmosphere and gradually settle on the Earth's surface. Scientists at the Bangladesh Atomic Energy Commission (BAEC) previously observed that Dhaka was the most Pb polluted city in the world for a part of 1996. A 17-month survey conducted by the scientists of BAEC detected 463 $ng \cdot m^{-3}$ Pb for $PM_{2.5}$ in Dhaka air during the dry months (between November 1994 and January 1996) [37]. However, the atmospheric Pb concentration in Dhaka is decreasing gradually, presumably due to the ban on leaded gasoline in Bangladesh, although it is still higher than that found in European cities and Far East Asian sites Taiwan (133 $ng \cdot m^{-3}$) [38]. However, it is lower than that reported for Southeast Asian Sites such as Karachi, Pakistan, and Delhi, Mumbai, and Kanpur, India [39].

Chromium

Chromium (Cr) is a grey, hard metal most commonly found in the trivalent state in nature. Hexavalent chromium compounds are also found in small quantities. Chromite ($FeOCr_2O_3$) is the only ore containing a significant amount of chromium. The ore has not been found in the pure form; its highest grade contains about 55% chromic oxide. Chromium is used in metal alloys and pigments for paints, cement, paper, rubber, and other materials. Due to the lot of tanneries in the vicinity of Dhaka city, chromium is relatively higher in Dhaka city. The concentration of chromium found in this study was 9.2 $ng \cdot m^{-3}$ varying from 4.4 to 15.2 $ng \cdot m^{-3}$ (Table 1 and Table 2). Information on concentrations of total and speciated chromium in the atmosphere is limited. Measure ments carried out above the North Atlantic, north of latitude 30° N, several thousands of kilometers from major land masses, showed concentrations of chromium of 0.07 - 1.1 ng·m^{-3} [40]. The concentrations above the South Pole were slightly lower. The following chromium concentrations have also been reported: 0.7 ng·m^{-3} in the Shetland Islands and Norway, 0.6 ng·m^{-3} in northwest Canada, 1 - 140 ng·m^{-3} in continental Europe, 20 - 70 ng·m^{-3} in Japan, and 45 - 67 ng·m^{-3} in Hawaii [41].

Table 2. Comparison of the average concentration of the determined trace metals with other locations of the world.

Metals	Current study	China[a]	India[b]	USA[c]	Japan[d]	Pakistan[e]	Taejon, Korea[f]	Birmingham, UK[g]
As	3.06	-	-	-	-	-	-	-
Cd	6.28	-	6.7	0.77	0.45	2	3.24	0.5
Ni	3.77	51	97	25.28	-	9	37.9	2
Cu	11.98	-	-	14.60	18.0	-	-	-
Pb	305.55	46	380	15.04	5.75	210	243	27
Cr	9.2	-	104	70.17	-	42	25.1	-
Fe	2057.0	5100	5220	512.4	869.3	930	1633	204
Mn	42.2	1210	-	1.89	34.7	57.0	50.3	6
Zn	303.3	274	-	2.99	1386.2	542	240	30
Sb	5.75	-	-	-	-	-	-	-
Se	2.43	-	-	-	-	-	-	-

[a]Mori et al., 2003; [b]Khillare et al., 2004; [c]Goforth et al., 2006; [d]Wanga et al., 2005; [e]Shaheen et al. 2005; [f]Kim et al. 2002; [g]Harrison et al. 1996.

Monitoring of the ambient air during the period of 1977- 1980 in many urban and rural areas of the United States of America showed chromium concentrations ranged from 5.2 ng·m^{-3} (24-hour background level) to 156.8 ng·m^{-3} (urban annual average) with the maximum concentration was about 684 ng·m^{-3} (24-hour average). Ranges of chromium levels in Member States of the European Union were given in a survey as follows: remote areas 0 - 3 ng·m^{-3}; urban areas: 4 - 70 ng·m^{-3}, and industrial areas 5 - 200 ng·m^{-3}. From the point of view of toxicity and carcinogenicity, chromium (VI) compounds are of much greater significance for workers and the general population than are trivalent and other valence states of chromium compounds. Therefore, chromium (VI) and chromium (III) have to be considered separately. This is, however, difficult to do when only total chromium is measured.

Iron (Fe)

Mining activities and other geochemical processes often result in the generation of acid mine drainage (AMD), a phenomenon commonly associated with mining activities. It is generated when pyrite (FeS_2) and other sulphide minerals in the aquifer and present and former mining sites are exposed to air and water in the presence of oxidizing bacteria, such as Thiobacillusferrooxidans, and oxidised to produce metal ions, sulphate and acidity. The average Fe concentration found in this study is 2057.0 $ng \cdot m^{-3}$ varying from 1242 to 3207 $ng \cdot m^{-3}$ (Table 1). Iron exhibited relatively higher values because Fe may originate from soil dusts, poorly managed transport, and building construction, among others, the major sources of Fe are both anthropogenic and crustal in origin, including iron and steel manufacturing units and the weathering of exposed Fe in urban areas.

Manganese (Mn)

Manganese (Mn) is an element widely distributed in the earth's crust. It is considered to bethe twelfth most abundant element and the fifth most abundant metal. Manganese does not occur naturally in a pure state; oxides, carbonates and silicates are the most important manganese-containing minerals. The most common manganese mineral is pyrolusite (MnO_2), usually mined in sedimentary deposits by open-cast techniques. Manganese occurs in most iron ores. Its content in coal ranges from 6.0 to 100 μgg^{-1}; it is also present in crude oil, but at substantially lowers concentrations. The average concentration of Mn found in this study was 42.2 $ng \cdot m^{-3}$ varying from 22.2 to 65.1 $ng \cdot m^{-3}$ (Table 1), which is slightly higher than the average concentration found in some developed countries of the world. In the Federal Republic of Germany, annual mean concentrations of manganese ranged between 3 and 16 $ng \cdot m^{-3}$ in Frankfurt, Main and Munich [42] ; in Belgium over the period 1972-1977, annual mean manganese concentrations of between 42 $ng \cdot m^{-3}$ and 456 $ng \cdot m^{-3}$ were reported [43]. The Environmental Agency of Japan reported an annual mean manganese concentration of about 20 - 800 $ng \cdot m^{-3}$ in Japanese cities, with maximum 24-hour concentrations of 2 - 3 μgm^{-3} [44].

Zinc (Zn)

Zinc is an element commonly found in the Earth's crust. It is released to the environment from both natural and anthropogenic sources; however, releases from anthropogenic sources are greater than those from natural sources. The primary anthropogenic sources of zinc in the environment (air, water, soil) are related to mining and metallurgic operations involving zinc and use of commercial products containing zinc. The average value of Zn was 303.3 ng·m^{-3} varying from 88.1 to 497.7 ng·m^{-3} (Table 1), which is much lower than the value (801 ng·m^{-3}) obtained at a previous measurement [21] for SPM. Elevated concentrations of Zn have also been observed for other Southeast Asian sites. The Zn concentration at Dhaka was much lower than that at Delhi [39] and Kanpur [45], India, but much higher than that found in Norway [46].

Antimony

Antimony is a silvery-white metal. Antimony oxide is used to produce fire retardants, paints, ceramics and fireworks. Antimony is released to the environment from natural sources and from its industrial processing. In the air, antimony is attached to very small particles. It may stay in the air for many days. The average concentration of Antimony found 5.47 ng·m^{-3} varying from 1.5 to 10.7 ng·m^{-3} (Table 1), which is slightly lower than the average value found in various cities of USA and higher than United Kingdom. In Chicago, the concentrations of antimony in air ranging from 1.4 to 55 ng·m^{-3} with an average value of 32 ng·m^{-3} whereas somewhat lower levels (0.4 to 4 ng·m^{-3}) were reported from seven different sites in the United Kingdom [47].

Selenium

Selenium is commonly found as selenide minerals in rocks that also contain sulphides of silver, copper, lead, and nickel—and in other forms in water and dry soils. Selenium is used in electronic and photographic equipment, glass, pigments, rubber, pesticides, dietary supplements, and livestock and poultry feed. Trace amounts of selenium are present in coal and oil. When electric utilities burn these fuels at their power plants, selenium is released in very small amounts. Selenium is released into the air by soils as they erode in wind. Breathing large amounts of selenium dust can irritate the lungs, and cause headaches and dizziness. In this study the average concentration of Selenium was found 2.43 ng·m^{-3} varying from 1.2 to 3.4 ng·m^{-3} (Table 1). It is satisfactory to say that the level of selenium in the air of Dhaka city is very relatively low. There is no previous measurement for selenium in atmosphere of Dhaka so far as we know.

Comparison of the Trace Metals in Dhaka with Other Cities of the World

The highest average concentrations for the measured trace elements were observed for iron, followed by lead, Zinc, Manganese, copper, Chromium, Cadmium, Antimony, Nickel, Arsenic, Selenium. Compared with other South Asian cities the determined concentrations of trace elements at urban Dhaka exhibited typically lower levels for some trace elements than observed in Lahore, Pakistan and Calcutta, India but for some trace elements it exhibited much higher values (Table 2). The average arsenic concentration in Dhaka is 3.06 $ng \cdot m^{-3}$ below the orientation value of the German TA-Luft of 13 $ng \cdot m^{-3}$ (German TA-Luft). Still, the average arsenic concentration in Dhaka is lower than at other south Asian polluted cities such as, Lahore, Mumbai and Calcutta. Average cadmium concentrations were close to the WHO [19] guideline value (5 $ng \cdot m^{-3}$) for the annual mean. The trace metal concentration in Dhaka is relatively low compared than other south East Asian cities—maybe there are not many metal processing industries in and around Dhaka city. But the concentration levels may be higher at other cities in Bangladesh with many metal and steel manufacturing industries e.g., Chittagong.

Correlation Factor

Correlation coefficient is a measure of the linear correlation between two variables and giving a value between +1 and −1. Where 1 is total positive correlation, 0 is no correlation, and −1 is the total negative correlation. It is widely used in the sciences as a measure of the degree of linear dependences between two variables. However, the correlation factor for various metals were derived and shown in the Table 3. As for the metal to metal corre- lation Manganese is highly correlated with Nickel ($r^2 = 0.84$), Iron ($r^2 = 0.97$) and Copper ($r^2 = 0.86$). Zinc is correlated with Cadmium ($r^2 = 0.77$), Chromium with Zinc ($r^2 = 0.65$), Iron with Nickel ($r^2 = 0.84$), Antimony with Arsenic ($r^2 = 0.92$), Selenium with Arsenic ($r^2 = 0.73$) and Antimony ($r^2 = 0.66$), Lead with Arsenic ($r^2 = 0.79$) and Antimony ($r^2 = 0.78$) indicating a common source of these metals as also manifested in strong correlations.

Table 3. Correlation coefficient of various metals with metals ($r^2 \geq 0.65$ is in bold).

	As	Cd	Zn	Ni	Cr	Fe	Sb	Se	**Mn**	**Pb**	**Cu**
As	1	0.08	0.26	0.22	−0.04	0.00	0.92	0.73	0.14	0.79	0.38
Cd		1.00	0.77	0.22	0.32	0.05	−0.09	0.52	0.24	0.49	0.50
Zn			1.00	0.26	0.65	0.26	0.01	0.66	0.44	0.39	0.57

Ni				1.00	0.51	0.84	−0.09	0.45	0.84	0.12	0.81
Cr					1.00	0.65	−0.29	0.43	0.43	−0.03	0.46
Fe						1.00	−0.28	0.18	0.97	−0.07	0.77
Sb							1.00	0.48	−0.16	0.78	0.04
Se								1.00	0.32	0.63	0.59
Mn									1.00	0.14	0.86
Pb										1.00	0.39
Cu											1

Enrichment Factor (EF)

Iron is a major constituent in the earth crust but also immaterial derived from the crust such as soil dust, road dust, dust from construction activities, etc., and also coal fly ash. The relative amount of Fe in the earth crusts is 5.0% m/m [48]. In dust from soil, shale, and other crust-related material the Fe content is not far from the crustal abundance. Also the inorganic material in coal is crust-related. The average Fe content in coal fly ash (total emitted dust) is around 8.5% m/m [49]. To roughly separate trace elements from crustal and non-crustal sources, crustal enrichment factors (EFs) were applied to the results. In Table 3 the enrichment factors were derived for the Dhaka urban aerosol (averages) relative to the composition of the earth crust and of coal fly ash using Fe as the reference element. The enrichment factor (EF) [50] is defined as the double ratio of the concentration of the determined element (X) to that of Fe in aerosol and in the reference sources e.g. earth crust or coal fly ash.

$$[50] \quad EF = \left(X/Fe \right)_{aerosol} / \left(X/Fe \right)_{crustor\ coal\ fly\ ash}$$

By convention, an arbitrary average EF value < 10 indicates that a trace element in particulate matter has a significant crustal source, and in contrast, an EF value of >10 is considered as a significant proportion of an element with a non-crustal source [51]. The enrichment factors of trace element with respect to the levels of the crustal enrichments are presented in Table 4. Cd, Pb and Zn are crustal elements (EF < 10) and others are non-crustal elements (EF > 10). Non-crustal elements associated with particulate matter near the highways mostly come from vehicle emissions [52] [53]. Concentrations of some trace elements (Cd, Pb, Zn) are considerably enriched in relation to the crustal abundances. The combustion of coal is the source for Cd, Cu, Pb and Zn, while tire rubber abrasion and brakes are found to be major contributors to Cd, Pb, Cu and Zn in the ambient air near high-traffic zones [54]. Zn could come from multiple sources, such as vehicle emissions, tire tread [55] [56], diesel soot, oil industries and coal combustion [57]. However, similarities are found between the abundance of most elements relative to coal fly ash. In particular EFs for the aerosol relative to coal fly ash for Cd, Cu, Pb, Zn, are not too far from one (within a factor of around 5, Table 4). The EF for As is 4 and for Cu is 5 relative to earth crust, and for coal 0 and 2 pointing to a potential geogenic as well as anthropogenic source. The highest enrichment factor for Cd, Pb, Zn showed that they are not from geogenic sources but they really introduced into the environment from anthropogenic sources such as coal fly ash.

CONCLUSION

People from all over the world are concerned more about the air pollution aspects due to the increased rate of mortality and morbidity and also multifarious affects of particulate pollution on our environment. In this regard it is imperative to have a systematic study ascertaining the facts concerning the nature, sources, and trends of the particulate pollution in city Dhaka. This study was undertaken especially to look into these aspects. During the study period it was found that the average concentration of some metals viz. Fe, Pb, Zn, Cu to be of higher order can most likely be attributed to rapid development, increased vehicle emissions to the atmosphere, and to the lack of sophisticated management of wastes and effluents from factories.

Table 4. Enrichment factor [50] analysis for the selected trace metals in Dhaka, Bangladesh.

Elements	Dhaka ng·m^{-3}	Earth crust mg·g^{-1}	Coal fly ash mg·g^{-1}	Dhaka EF/CRUST	Dhaka EF/FLY
As	3.06	0.018	0.400	4	0
Cd	6.28	0.00008	0.000	1908	0.0
Pb	305.6	0.013	2.200	571	6
Cu	11.98	0.055	0.300	5	2
Fe	2057	50.00	85.00	1	1
Zn	303.3	0.070	2.300	105	5

Enrichment factors "EF" (reference element Fe) for Dhaka urban aerosol relative to earth crust and to coal fly ash.

Although for some metals like Ni, Cu, Se, Sb contamination is not as severe as in other developed cities of the world. Metals like Fe, Zn, Se, Pb, Cu showed strong inter correlations indicating isogenic source in nature, whereas As, Cd, Ni are neither correlated with any metal indicating different sources of emission. Enrichment factor analysis also enabled to identify the actual sources for the metals. Enrichment factors based on Fe as a reference element indicate that coal combustion is a major contributor to Fe, Cu, Pb and Zn in the Dhaka particulate matters. The comparison study also presents an alarming picture of airborne trace metals in Dhaka city. Therefore, it is high time to develop an air pollution abatement strategy to protect people from the hazardous effects arising from elevated atmospheric trace metal levels by the systematic study of air pollution.

ACKNOWLEDGEMENTS

Authors acknowledge the Department of Environment Sciences Informatics and Statistics, CàFoscari University of Venice, Italy for helping with chemical analysis. Authors also acknowledge the help of Md. Halim during Sampling.

REFERENCES

1.   Tsai, Y.I. and Cheng, M.T. (2004) Characterization of Chemical Species in Atmospheric Aerosols in a Metropolitan Basin. Chemosphere, 54, 1171-1181.http://dx.doi.org/10.1016/j.chemosphere.2003.09.021

2.   Park, S.S. and Kim, Y.J. (2005) Source Contributions to Fine Particulate Matter in an Urban Atmosphere. Chemosphere, 59, 217-226.http://dx.doi.org/10.1016/j.chemosphere.2004.11.001

3.   Watson, J.G., Zhu, T., Chow, J.C., Engelbrecht, J., Fujita, E.M. and Wilson, W.E. (2002) Receptor Modelling Application Framework for Particle Source Apportionment. Chemosphere, 49, 1093-1136. http://dx.doi.org/10.1016/S0045-6535(02)00243-6

4.   Quiterio, S.L., da Silva, C.R.S., Arbilla, G. and Escaleira, V. (2004) Metals in Airborne Particulate Matter in the Industrial District of Santa Cruz, Rio de Janeiro, in an Annual Period. Atmospheric Environment, 38, 321-331.http://dx.doi.org/10.1016/j.atmosenv.2003.09.017

5.   Arditsoglou, A. and Samara, C. (2005) Levels of Total Suspended Particulate Matter and Major Trace Elements in Kosovo: A Source Identification and Apportionment Study. Chemosphere, 59, 669-678. http://dx.doi.org/10.1016/j.chemosphere.2004.10.056

6.   Shah, M.H. (2009) Atmospheric Particulate Matter: Trace Metals and Size Fractionation. VDM Verlag Dr. Muller, Saarbrucken, 228.

7.   Shaheen, N., Shah, M.H. and Jaffar, M. (2005) A Study of Airborne Selected Metals and Particle Size Distribution in Relation to Climatic Variables and Their Source Identification. Water, Air, and Soil Pollution, 164, 275-294. http://dx.doi.org/10.1007/s11270-005-3542-1

8.   Callen, M.S., de la Cruz, M.T., Lopez, J.M., Navarro, M.V. and Mastral, A.M. (2009) Comparison of Receptor Models for Source Apportionment of the PM_{10} in Zaragoza (Spain). Chemosphere, 76, 1120-1129.http://dx.doi.org/10.1016/j.chemosphere.2009.04.015

9.   Prieditis, H. and Adamson, I.Y.R. (2002) Comparative Pulmonary Toxicity of Various Soluble Metals Found in Urban Particulate Dusts. Experimental Lung Research, 28, 563-576. http://dx.doi.org/10.1080/01902140290096782

10.   Magas, O.K., Gunter, J.T. and Regens, J.L. (2007) Ambient Air Pollution and Daily Pediatric Hospitalizations for Asthma. Environmental Science and Pollution Research, 14, 19-23. http://dx.doi.org/10.1065/espr2006.08.333

11.   Wild, P., Bourgkard, E. and Paris, C. (2009) Lung Cancer and Exposure to Metals: The Epidemiological Evidence. Method Molecular Biology, 472, 139-167.http://dx.doi.org/10.1007/978-1-60327-492-0_6

12.   Dreher, K.L., Jaskot, R.H., Lehmann, J.R., Richards, J.H., McGee, J.K., Ghio, A.J. and Costa, D.L. (1997) Soluble Transition Metals Mediate Residualoil Fly Ash Induced Acute Lung Injury. Journal of Toxicology Environmental Health, 50, 285-305.

13.   Schaumann, F., Borm, P.J., Herbrich, A., Knoch, J., Pitz, M., Schins, R.P., Luettig, B., Hohlfeld, J.M., Heinrich, J. and Krug, N. (2004) Metal-Rich Ambient Particles (Particulate Matter 2.5) Cause Airway Inflammation in Healthy Subjects. American Journal of Respiratory and Critical Care Medicine, 170, 898-903.http://dx.doi.org/10.1164/rccm.200403-423OC

14.   Schwarze, P.E., Totlandsdal, A., Herseth, I.J., Holme, A.J., Låg, M., Refsnes, M., Øvrevik, J., Sandberg, W. and Bøl- ling, K.A. (2010) Importance of Components and Sources for Health Effects of Particulate Air Pollution. Vanda Villanyi.

15.   Gurgueira, S.A., Lawrence, J., Coull, B., Murthy, G.G. and Gonzalez-Flecha, B. (2002) Rapid Increases in the Steady- State Concentration of Reactive Oxygen Species in the Lungs and Heart after Particulate Air Pollution Inhalation. Environmental Health Perspective, 110, 749-755. http://dx.doi.org/10.1289/ehp.02110749

16.   Bell, L.M., Ebisu, K., Peng, D.R., Samet, M.J. and Dominici, F. (2009) Hospital Admissions and Chemical Composition of Fine Particle Air

Pollution. American Journal of Respiratory and Critical Care Medicine, 179, 1115-1120. http://dx.doi.org/10.1164/rccm.200808-1240OC

17. Fiaz, A. and Strum, P.J. (2000) New Directions: Air Pollutions and Road Traffic in Developing Countries. Atmospheric Environment, 34, 4745-4746.http://dx.doi.org/10.1016/S1352-2310(00)00255-7

18. Salam, A., Al Mamoon, H., Ullah, M.B. and Ullah, S.M. (2012) Measurement of the Atmospheric Aerosol Particle Size Distribution in a Highly Polluted Mega-City in Southeast Asia (Dhaka-Bangladesh). Atmospheric Environment, 59, 338-343.http://dx.doi.org/10.1016/j.atmosenv.2012.05.024

19. WHO (2000) Cadmium. Air Quality Guidelines for Europe, 2nd Edition, World Health Organization Regional Office for Europe, Copenhagen.

20. Rahman, M.A., Rahim, A., Siddique, N.A. and Alam, A.M.S. (2013) Studies on Selected Metals and Other Pollutants in Urban Atmosphere in Dhaka Bangladesh. Dhaka University Journal of Science, 61, 41-46. http://dx.doi.org/10.3329/dujs.v61i1.15094

21. Salam, A., Bauer, H., Kassin, K., Ullah, S.M. and Puxbaum, H. (2003) Aerosol Chemical Characteristics of a Mega- City in Southeast Asia (Dhaka-Bangladesh). Atmospheric Environment, 37, 2517-2528. http://dx.doi.org/10.1016/S1352-2310(03)00135-3

22. Salam, A., Ullah, M.B., Islam, M.D., Salam, M.A. and Ullah, S.M. (2011) Carbonaceous Species in Total Suspended Particulate Matters at Different Urban and Suburban Locations in the Greater Dhaka Region, Bangladesh. Air Quality, Atmosphere and Health, 6, 239-245. http://dx.doi.org/10.1007/s11869-011-0166-z

23. CEN (2005) Ambient Air Quality-Standard Gravimetric Measurement Method for the Determination of the $PM_{2.5}$ Mass Fraction of Suspended Particulate Matter. EN 14907:2005.

24. Mondol, M.N., Khaled, M., Chamon, A.S. and Ullah, S.M. (2014) Trace Metal Concentration in Atmospheric Aerosols in Some City Areas of Bangladesh. Bangladesh Journal of Scientific and Industrial Research, 49, 263-270.

25. Huang, X.D., Olmez, I., Aras, N.K. and Gordon, G.E. (1994) Emissions of Trace Elements from Motor Vehicles: Potential Marker Elements and Source Composition Profile. Atmospheric Environment, 28, 1385-1391. http://dx.doi.org/10.1016/1352-2310(94)90201-1

26. Chuersuwan, N., Nimrat, S., Lekphet, S. and Kerdkumrai, T. (2008) Levels and Major Sources of $PM_{2.5}$ and PM_{10} in Bangkok Metropolitan Region. Environment International, 34, 671-677.

27. Sharma, V.K. and Patil, R.S. (1992) Chemical Composition and Source Identification of Bombay Aerosol. Environmental Technology, 13, 1043-1052.http://dx.doi.org/10.1080/09593339209385241

28. Salam, A., Hossain, T., Siddique, M.N.A. and Alam, A.M.S. (2008) Characteristics of Atmospheric Trace Gases, Particulate Matter, and Heavy Metal Pollution in Dhaka, Bangladesh. Air Quality Atmosphere Health, 1, 101-109.http://dx.doi.org/10.1007/s11869-008-0017-8

29. Chan, W.H and Lusis, M.A. (1986) Smelting Operations and Trace Metals in Air and Precipitation in the Sudbury Basin. Advances in Environmental Science and Technology, 17, 113-143.

30. Norseth, T. (1994) Environmental Pollution around Nickel Smelters in the Kola Peninsula (Russia). Science of the Total Environment, 148, 103-108.http://dx.doi.org/10.1016/0048-9697(94)90389-1

31. Sivertsen, B. (1991) Air Pollution in the Border Areas of Norway/Soviet Union January 1990-March 1991. Lillestrøm, Norwegian Institute for Air Research, (NILU OR: 69/91; Ref:0-8976). Analytik, Institut fur Anorganische und AngewandteChemie, Universit at Hamburg, Germany.

32. Saltzman, B.E., Cholak, J. and Schafer, L.J. (1985) Concentrations of Six Metals in the Air of Eight Cities. Environmental Science and Technology, 19, 328-333.http://dx.doi.org/10.1021/es00134a004

33. Schmidt, J.A. and Andren, A.W. (1990) The Atmospheric Chemistry of Nickel. In: Nriagu, J.O., Ed., Nickel in the Environment, Vol. 1999, Wiley, New York, 93-137.

34. Bennett, B.J. (1994) Environmental Nickel Pathways to Man. In: Sunderman Jr., F.W., Ed., Nickel in the Human Environment, Lyon International Agency for Research on Cancer, Lyon, 487-495.

35. Smith, D.J.T., Harrison, R.M., Luhana, L., Pio, C.A., Castro, L.M., Tariq, M.N., Hayat, S. and Quraishi, T. (1996) Concentrations of Particulate Airborne Polycyclic Aromatic Hydrocarbons and Metals Collected in Lahore, Pakistan. Atmospheric Environment, 30, 4031-4040. http://dx.doi.org/10.1016/1352-2310(96)00107-0

36. Mori, I., Nishikawa, M., Tanimura, T. and Quan, H. (2003) Change in Size Distribution and Chemical Composition of Kosa (Asian Dust) Aerosol during Long Range Transport. Atmospheric Environment, 37, 4253-4263. http://dx.doi.org/10.1016/S1352-2310(03)00535-1

37. Khaliquzzaman, M., Biswas, S.K., Tarafdar, S.A., Islam, A. and Khan, A.H. (1997) Trace Element Composition of Size Fractionated Airborne Particulate Matter in Urban and Rural Areas in Bangladesh. Report,

Accelerator Facilities Division and Chemistry Division, Atomic Energy Centre, Dhaka.

38. Fung, Y.S. and Wong, L.W.Y. (1995) Apportionment of Air Pollution Sources by Receptor Models in Hong Kong. At- mospheric Environment, 29, 2041-2048.http://dx.doi.org/10.1016/1352-2310(94)00239-H

39. Balachandran, S., Meena, B.R. and Khillare, P.S. (2000) Particle Size Distribution and Its Elemental Composition in the Ambient Air of Delhi. Environment International, 26, 49-54.http://dx.doi.org/10.1016/S0160-4120(00)00077-5

40. Duce, R.A., Hoffman, G.L. and Zoller, W.H. (1975) Atmospheric Trace Metals at Remote Northern and Southern He- misphere sites. Pollution or Natural Science, 187, 59-61.

41. Bowen, H.J.M. (1979) Environmental Chemistry of the Elements. Academic Press, London.

42. Georgii, H.W. and Müller, J. (2004) Schwermetallaerosole in der Großstadtluft (Heavy Metal). In: Godish, T., Ed., Air Quality, 4th Edition.

43. Kretzschmar, J.G., Delespaul, I. and De Rijck, T. (1980) Heavy Metal Levels in Belgium: A Five Year Survey. Science of the Total environment, 14, 85-97.http://dx.doi.org/10.1016/0048-9697(80)90128-X

44. Wang, X.L., Sato, T., Xing, B.S., Tamamura, S. and Tao, S. (2005) Source Identification, Size Distribution and Indica- tor Screening of Airborne Trace Metals in Kanazawa, Japan. Aerosol Science, 36, 197-210. http://dx.doi.org/10.1016/j.jaerosci.2004.08.005

45. Sharma, M. and Maloo, S. (2005) Assessment of Ambient Air PM_{10} and $PM_{2.5}$ and Characterization of PM_{10} in the City of Kanpur, India. Atmospheric Environment, 39, 6015-6026. http://dx.doi.org/10.1016/j.atmosenv.2005.04.041

46. NILU (Norwegian Institute of Air Research) (2002) Heavy Metals and POPS within the EMEP Region 2000. Report -EMEP/CCC-9/ 2002, Norwegian Institute of Air Research, Kjeller.

47. Goforth, M.R. and Christoforou, C.S. (2006) Particle Size Distribution and Atmospheric Metals Measurements in a Rural Area in the South Eastern USA. Science of the Total Environment, 217-227. http://dx.doi.org/10.1016/j.scitotenv.2005.03.017

48. Mason, B. and Moore, C. (1982) Principles of Geochemistry. Wiley, New York.

49. Steiger, M. (1991) Die anthropogenen und natürlichen Quellenurbaner und mariner Aerosol charakterisiert und quantifiziertdurch

Multielementanalyse und chemische Receptormodelle, Schriftenreihe Angewandte.

50. Puxbaum, H. (1993) Luftchemie. Schriftenreihe "Modern Analytical Chemie," Band 5. Institutfür Analytical Chemie, Technische Universät Wien.

51. Chester, R., Nimmo, M. and Preston, M.R. (1999) The Trace Metal Chemistry of Atmospheric Dry Deposition Samples Collected at Cap Ferrat: A Coastal Site in the Western Mediterranean. Marine Chemistry, 68, 15-30. http://dx.doi.org/10.1016/S0304-4203(99)00062-6

52. Weckwerth, G. (2001) Verification of Traffic Emitted Aerosol Components in the Ambient Air of Cologne (Germany). Atmospheric Environment, 35, 5525-5536.http://dx.doi.org/10.1016/S1352-2310(01)00234-5

53. Sternbeck, J., Sjodin, A.A. and Andersson, K. (2002) Metal Emissions from Road Traffic and the Influence of Resuspension-Results from Two Tunnel Studies. Atmospheric Environment, 36, 4735-4744. http://dx.doi.org/10.1016/S1352-2310(02)00561-7

54. Singh, M., Jaques, P.A. and Sioutas, C. (2002) Size Distribution and Diurnal Characteristics of Particle-Bound Metals in Source and Receptor Sites of the Los Angeleles Basin. Atmospheric Environment, 36, 1675-1689. http://dx.doi.org/10.1016/S1352-2310(02)00166-8

55. Rogge, W.F., Hildemann, L.M., Mazurek, M.A., Cass, G.R. and Simoneit, B.R.T. (1993) Sources of Fine Organic Ae- rosols: 3 Road Dust, Tire Debris, and Organometallic Brake Lining Dust: Roads as Sources and Sinks. Environmental Science and Technology, 27, 1892-1904. http://dx.doi.org/10.1021/es00046a019

56. Nriagu, J.O. and Pacyna, J.M. (1988) Quantitative Assessment of Worldwide Contamination of Air, Water and Soils by Trace Metals. Nature, 333, 134-199.http://dx.doi.org/10.1038/333134a0

57. Pacyna, J.M. (1998) Source Inventories for Atmospheric Trace Metals. In: Harrison, R.M. and van Grieken, R.E., Eds., Atmospheric Particles, IUPAC Series on Analytical and Physical Chemistry of Environmental Systems, Vol. 5, Wiley, Chichester, 385-423.

Chapter 11

THE INFLUENCE OF EDDY DIFFUSIVITY VARIATION ON THE ATMOSPHERIC DIFFUSION EQUATION

A. A. Marrouf[1], Khaled S. M. Essa[1], Maha S. El-Otaify[1], Adel S. Mohamed[2], Galal Ismail[2]

[1]Department of Mathematics and Theoretical Physics, Atomic Energy Authority, Cairo, Egypt

[2]Department of Mathematics, Faculty of Science, Zagazig University, Zagazig, Egypt

ABSTRACT

The advection diffusion equation was solved analytically using separation of variables technique, considering first the wind speed and eddy diffusivity as constants; second as variables dependent on vertical height z. Comparison between predicted two models and observed concentration on Inshas, Cairo (Egypt) is done.

INTRODUCTION

Air pollutants released from various sources affect directly or indirectly man and his environment. Air pollutants emitted from different sources are transported dispersed or deposited my meteorological and topographical conditions. Dispersion of pollutants in the atmosphere is governed by the following dominant mechanisms [1], mean air flow that transports the pollutants downwind and turbulent velocity fluctuations that disperse the pollutants in all directions. Under moderate to strong winds, the continuously emitted pollutants from a cone- shaped plume in the downwind direction of the source. In this case, advection in the mean wind direction dominates over diffusion and dispersion in the crosswind and vertical directions is assumed to be non-Gaussian. Along-wind diffusion is particularly important near the leading edge of the plume, where uncontaminated fluid from upwind mixes with the mass initially released [2].

Analytical solutions of the advection-diffusion equation are usually obtained just for stationary conditions and by making strong assumptions about the eddy diffusivity coefficients (K) and wind speed profiles (U). They are assumed as constant throughout the whole Atmospheric Boundary Layer (ABL) or follow a power law [3] - [6]. Moreira et al. presented a solution of the advection-diffusion equation based on the Laplace transform considering the ABL as a multilayer system [7]. Number of dispersion regulatory models includes improved dispersion algorithms in terms of fundamental scaling parameters [8] - [11]. Gryning et al. suggested a modeling approach composed by individual models [12] ; each one based the specific turbulent structure of the regimes in the ABL, following [13]. The models give the crosswind-integrated concentrations at the ground, for non-buoyant releases from a continuous point source. They are limited to horizontally homogeneous conditions and travel distances less than 10 km.

Palazzi et al. have proposed a simple model for studying the diffusion of substances emitted in steady-state releases of short duration assuming the presence of an infinite mixing layer [14]. The Gaussian models, which are the best known and most widely used, are based on a solution of the two-dimensional advection equation where both the wind and exchange coefficients are assumed to be constant. The Gaussian model solution is forced to represent an inhomogeneous atmosphere through empirical dispersion parameters [15].

In this study, we have formulated a mathematical model for dispersion of air pollutants in moderated winds by taking into account the diffusion in vertical height direction and advection along the mean wind. The eddy diffusivity and wind speed are assumed to be constant. An analytical solution has been obtained for the resulting advection-diffusion equation with the physically relevant boundary conditions. The moderate data collected during the convective conditions. Nine experiments were conducted at Inshas site, Cairo-Egypt [16], which used to investigate the analytical solution.

MATHEMATICAL TREATMENT

The dispersion of pollutants in the atmosphere is governed by the basic atmospheric diffusion equation. Under the assumption of incompressible flow, atmospheric diffusion equation based on the Gradient transport theory can be written in the rectangular coordinate system as:

$$\frac{\partial C}{\partial t}+u\frac{\partial C}{\partial x}+v\frac{\partial C}{\partial y}+w\frac{\partial C}{\partial z}=\frac{\partial}{\partial x}\left(K_x\frac{\partial C}{\partial x}\right)+\frac{\partial}{\partial y}\left(K_y\frac{\partial C}{\partial y}\right)+\frac{\partial}{\partial z}\left(K_z\frac{\partial C}{\partial z}\right)+S \tag{1}$$

where C is the mean concentration of a pollutant (Bq/m³), (µg/m³) and (ppm); S is the source term, respectively; (u, v, w) and (k_x, k_y, k_z) are the components of wind and diffusivity vectors in x, y and z directions, respectively, in an Eulerian frame of reference.

The following assumptions are made in order to simplify Equation (1):

- Steady-state conditions are considered, i.e. $\partial C/\partial t = 0$.
- As the vertical velocity is much smaller than the horizontal one in x-direction, the term $w(\partial C/\partial z)$ is neglected.
- x-axis is oriented in the direction of mean wind u = U and U much greater than the wind speed V in y-direction the term $v(\partial C/\partial y)$ is neglected).
- Source (physical/chemical) pollutants are ignored so that S = 0.

With the above assumptions, Equation (1) reduces to:

$$U\frac{\partial C}{\partial x} = \frac{\partial}{\partial x}\left(K_x\frac{\partial C}{\partial x}\right) + \frac{\partial}{\partial y}\left(K_y\frac{\partial C}{\partial y}\right) + \frac{\partial}{\partial z}\left(K_z\frac{\partial C}{\partial z}\right).$$

(2)

The advection term in x direction is larger than the diffusion in x direction then we will neglect the diffusion term in x direction,

$$U\frac{\partial C}{\partial x} = \frac{\partial}{\partial y}\left(K_y\frac{\partial C}{\partial y}\right) + \frac{\partial}{\partial z}\left(K_z\frac{\partial C}{\partial z}\right).$$

(3)

Equation (3) is solved together with the following boundary conditions.

- The is assumed to be a perfectly total absorption i.e.,

$$\frac{\partial C(x,z)}{\partial z} = 0 \quad \text{at} \quad z = 0, z = h.$$

(4)

- The pollutant is totally penetrate through the top of the inversion/ mixed layer located at height h, i.e.

$$C(x,y,z) = 0 \quad \text{at} \quad z = h.$$

(5)

- A continuous point source with strength Q is assumed to be located at the point $(0, y_s, z_s)$, i.e.

$$UC = Q\delta(z - z_s) \quad \text{as} \quad x = 0$$

(6)

where $\delta(\cdots)$ is Dirac's delta function.

- Far away from the source, the concentration decreases to zero, i.e.

$C \rightarrow 0$ as $xy, z \rightarrow \infty.$

$$(7)$$

Variable Eddy Diffusivity and Wind Speed

Here we will use Equation (3), considering the wind speed U as linear of z:

$$U = k_o u_* z, \ z \neq 0 \ \text{and} \ U = U_0 \ \text{at} \ z = 0$$

$$(8)$$

and eddy diffusivity k_z is expressed as functions of power law of z as:

$$k_z = u_1 z^n$$

$$(9)$$

where k_o is Von-Karmen constant and u_* is the friction velocity. Where u_1 is turbulence intensity.

Also after integrating Equation (3) with respect to y from $(-\infty$ to $\infty)$, Equation (2) becomes:

$$k_o u_* z \frac{\partial C_y}{\partial x} = \frac{\partial}{\partial z}\left(u_1 z^n \frac{\partial C_y}{\partial z} \right)$$

$$(10)$$

which is simply reads:

$$\frac{\partial C_y}{\partial x} = \frac{u_1}{k_o u_*} z^{n-1} \frac{\partial^2 C_y}{\partial z^2} + \frac{u_1 n}{k_o u_*} z^{n-2} \frac{\partial C_y}{\partial z}.$$

$$(11)$$

One can solve the two-dimensional partial differential Equation (11) analytically by using the separation of variables technique. We take the solution of Equation (11) of the form:

$$C_y(x, z) = X(x) \cdot Z(z).$$

$$(12)$$

Differentiating Equation (12) partially with respect to x and z and substituting in Equation (11), we get two ordinary differential equations in the variables X and Z as follows:

$$\frac{1}{X} \frac{dX}{dx} = -\lambda^2$$

$$(13)$$

and

$$\frac{\alpha z^{n-1}}{Z} \frac{d^2 Z}{dz^2} + \frac{\beta z^{n-2}}{Z} \frac{dZ}{dz} = -\lambda^2$$

$$(14)$$

where λ^2 is a constant, $\alpha = u_1/k_o u_*$ and $\beta = u_1 n/k_o u_*$.

The general solution of Equation (13) is given by

$$X(x) = \gamma e^{-\lambda^2 x}$$

$$(15)$$

where γ is a constant.

Equation (14) becomes:

$$z^2 \frac{d^2 Z}{dz^2} + nz \frac{dZ}{dz} + \frac{\lambda^2}{\alpha} z^{3-n} Z = 0.$$

(16)

Equation (16) which simply reads:

$$z_*^2 \frac{d^2 Z_*}{dz_*^2} + z_* \frac{dZ_*}{dz_*} + \left[\eta^2 z_*^2 - \mu^2 \right] Z_* = 0$$

(17)

where $\eta^2 = 4\lambda^2 / \alpha (3-n)^2$, $\mu = 1 - n/3 - i$.

The solution of Equation (14) is obtained in different boundary conditions as follows:

Equation (10) along with the following boundary condition corresponding to Equation (4) and Equation (5):

$$Z = 0 \quad \text{at} \quad z = 0, h.$$

(18)

On changing the dependent Z and independent z variables in Equation (16) by means of the substitutes:

$$Z = z_*^{\frac{1-n}{3-n}} Z_*$$

$$z_* = z^{\frac{3-n}{2}}$$

(19)

Equation (17) is a Bessel equation and has a solution [17] :

$$Z = z^{\frac{1-n}{2}} \left[A J_\mu \left(\eta z^{\frac{3-n}{2}} \right) + B J_{-\mu} \left(\eta z^{\frac{3-n}{2}} \right) \right]$$

(20)

where j_μ and $J_{-\mu}$ the Bessel functions of first kind of order μ and $-\mu$, respectively, A and B are constants, application of the boundary condition Equation (18) at z = 0 in Equation (20) yields B = 0 and condition z = h Equation (18) gives rise:

$$h^{\frac{1-n}{2}} J_\mu \left(\eta h^{\frac{3-n}{2}} \right) = 0.$$

(21)

Equation (21) this represents Storm-Liouville Eigen value problem which have the corresponding Eigen functions:

$$Z_\alpha(z) = z^{\frac{1-n}{2}} J_\mu\left(\eta_\alpha z^{\frac{3-n}{2}}\right) \quad \alpha = 1, 2, 3, \cdots, \infty.$$

(22)

The general of Equation (10) is obtained by using Equation (15), Equation (21) and Equation (22) as:

$$C_y(x,z) = z^{\frac{1-n}{2}} \left[\sum_{\alpha=1}^{\infty} A_\alpha J_\mu\left(\eta_\alpha z^{\frac{3-n}{2}}\right) \exp\left(-\lambda^2 x\right) \right]$$

(23)

where $A_\alpha \alpha = 1, 2, 3, \cdots, \infty$ are the unknown coefficients. Equation (23) represent the concentration distribution C_y through the Fourier-Bessel series [18] corresponding to a set of Eigen function Z_α.

Estimation of the coefficients A_α's for crosswind integrated concentrations: The source at x = 0, Equation (6) gives:

$$k_0 u_* z^{\frac{3-n}{2}} \left[\sum_{\alpha=1}^{\infty} A_\alpha J_\mu\left(\eta_\alpha z^{\frac{3-n}{2}}\right) \right] = Q_p \delta(z - z_s).$$

(24)

To determine the values of A_α we use the orthogonally of Eigen functions series [18].

Multiplying Equation (24) by $z^{\frac{1-n}{2}} J_\mu\left(\eta_\beta z^{\frac{3-n}{2}}\right)$ $\beta \geq 0$ and integrating according to z from 0 to h, we get:

$$A_\beta = \frac{2 Q_p z_s^{\frac{1-n}{2}}}{k_0 u_* h^2} * \frac{J_\mu\left(\eta_\beta z_s^{\frac{3-n}{2}}\right)}{J_{\mu+1}^2\left(\eta_\beta h^{\frac{3-n}{2}}\right)} \quad \beta \geq 1.$$

(25)

Substituting A_β in Equation (23), the final solution is given as follows:

$$C_y(x,z) = Q_p \frac{2(z z_s)^{\frac{1-n}{2}}}{k_0 u_* h^2} \sum_{\alpha=1}^{\infty} \frac{J_\mu\left(\eta_\alpha z^{\frac{3-n}{2}}\right) J_\mu\left(\eta_\alpha z_s^{\frac{3-n}{2}}\right)}{J_{\mu+1}^2\left(\eta_\beta h^{\frac{3-n}{2}}\right)} \exp\left(-\lambda^2 x\right).$$

(26)

In which $\eta_\beta h^{\frac{3-n}{2}}$ is given as:

$$J_\mu\left(\eta_\beta h^{\frac{3-n}{2}}\right) = 0.$$

$$(27)$$

Eddy Diffusivity and Wind Speed as Constant

Here we will use Equation (3), considering the wind speed U and eddy diffusivity k_z as constant:

Also after integrating Equation (3) with respect to y from ($-\infty$ to ∞), Equation (2) becomes:

$$u\frac{\partial C_y}{\partial x} = k\frac{\partial^2 C_y}{\partial z^2}$$

$$(28)$$

which is simply reads:

$$\frac{\partial C_y}{\partial x} = \frac{k}{u}\frac{\partial^2 C_y}{\partial z^2}.$$

$$(29)$$

One can solve the two-dimensional partial differential Equation (29) analytically by using the separation of variables technique. We take the solution of Equation (29) of the form:

$$C_y(x,z) = F(x)G(z).$$

$$(30)$$

Differentiating (30) partially with respect to x and z and substituting in Equation (29), we get two ordinary differential equations in the variables F(x) and G(x) as follows:

$$\frac{1}{F(x)}\frac{dF(x)}{dx} = -\lambda^2$$

$$(31)$$

and

$$\frac{k_z}{u}\frac{d^2G(z)}{dz^2} = -\lambda^2 G(z)$$

$$(32)$$

where λ^2 is a constant.

The general solution of Equation (31) is given by

$$F(x) = \gamma e^{-\lambda^2 x} \tag{33}$$

where γ is a constant.

Equation (32) becomes:

$$\frac{k_z}{u}\frac{d^2 G(z)}{dz^2} + \lambda^2 G(z) = 0 \tag{34}$$

which have solution

$$G(z) = A\cos(\lambda^2 z) + Bi\sin(\lambda^2 z) \text{ where } i^2 = -1 \tag{35}$$

where A and B are constant.

Then from Equation (33) and Equation (35) the general solution

$$C(x,z) = e^{-\lambda^2 x}\left(A\cos(\lambda^2 z) + Bi\sin(\lambda^2 z)\right). \tag{36}$$

By differentiate Equation (36) with respect to z and applying the boundary conditions we get:

$$\frac{\partial C(x,z)}{\partial z} = e^{-\lambda^2 x}\left(A\lambda^2 \sin(\lambda^2 z) + B\lambda^2 i\cos(\lambda^2 z)\right). \tag{37}$$

Appling the boundary condition Equation (4) on Equation (37) which gives $B = 0$ and Equation (36) becomes:

$$C(x,z) = A\cos(\lambda^2 z)e^{-\lambda^2 x}. \tag{38}$$

Again apply the boundary condition Equation (6) leads to

$$A = \frac{Q}{uz_s}\sec(\lambda^2 z_s). \tag{39}$$

Substituting A in Equation (38), the final solution is given as follows:

$$C(x,z) = \frac{Q}{uz_s}e^{-\lambda^2 x}\cos(\lambda^2 z)\sec(\lambda^2 z_s). \tag{40}$$

In the Previous section we used the wind speed and eddy diffusivity as functions in the vertical height z, and we had the solution Equation (26). Now we have two forms of the solutions Equation (26) and Equation (40).

APPLICATIONS

Source Data

The diffusion data for the estimating were gathered during ^{135}I isotope tracer nine experiments in moderate wind with unstable conditions at Inshas, Cairo. During each run, the tracer was released from source has height 43 m for twenty four hours working, where the air samples were collected during half hour at a height 0.7 m.

We collected air samples from 92 m to 184 m around the source in AEA, Egypt. The study area is at, dominated by sand soil with poor vegetation cover. The air samples collected were analyzed in Radiation Protection Department, NRC, AEA, Cairo, Egypt using a high volume air sampler with 220 V = 50 Hz bias [10]. Meteorological data have been provided by the measurements done at 10 and 60 m. Table 1 gives the data information about the diffusion tests and the wind vectors.

Table 1. Meteorological data of the nine convective test runs at Inshas site in March and May 2006.

Run No.	Work-ing hours	Release rate (Bq)	Wind speed $(m \cdot s^{-1})$	Wind direction (deg)	W_* $(m \cdot s^{-1})$	Z_i (m)	P-G stabil-ity class
1	48	1028571	4	301.1	2.27	600.85	A
2	49	1050000	4	278.7	3.05	801.13	A
3	1.5	42857.14	6	190.2	1.61	973	B
4	22	471428.6	4	197.9	1.23	888	C
5	23	492857.1	4	181.5	0.958	921	A
6	24	514285.7	4	347.3	1.3	443	D
7	28	1007143	4	330.8	1.51	1271	C
8	48.7	1043571	4	187.6	1.64	1842	C
9	48.25	1033929	4	141.7	2.1	1642	A

In addition, it contains values of vertical velocity scale (w_*) and mixing height (z_i). The data from these nine unstable test runs have been utilized for the following analysis.

Table 1 gives information about the diffusion tests and the wind vectors. In addition, it contains values of the vertical velocity scale (w_*).

Model Parameters

For the concentration computations, we require the knowledge of wind speed, wind direction, source strength, the dispersion parameters, mixing height and the vertical scale velocity. Wind speeds are greater than 3 m/s most of the time even at 10 m level. Further the variation wind direction with time is also visible. The analytical expressions depend upon downwind distance, vertical distance and atmospheric stability. The atmospheric stability has been calculated from Monin-Obukhov length scale (1/L) [19] based on friction velocity, temperature, and surface heat flux.

RESULTS AND DISCUSSION

The concentration is computed using data collected at vertical distance of a 30 m multi-level micrometeorological tower. In all a test runs were conducted for the purpose of computation. The concentration at a receptor can be computed in the following way:

Applying formula Equation (26) which contains the wind sped and eddy diffusivity as variable and Equation (40) which contains the wind sped and eddy diffusivity as constant at y = 0.0 for half hourly averaging.

Table 2 contains the observed concentrations Bq/m^3 and proposed concentrations in bounded and unbounded cases.

As an illustration, results computed from these approaches are shown in Table 2, for nine typical tests conducted at Inshas site, Cairo-Egypt [16]. This table shows that the predicted concentrations for [135]I using Equation (26) is very near to the observed concentration more than the predicted concentrations using Equation (40), because the eddy diffusivity and the wind speed were used as constants, on the other hand the eddy diffusivity and the wind speed had been used as functions in vertical height z, in Equation (26).

Figure 1 shows the variation of predicted and observed concentration of [135]I with the downwind distance. One gets good agreement between observed and predicted concentration Equation (26) more than predicted concentration Equation (40).

Figure 2 shows that the predicted concentrations which are estimated from Equation (26) and Equation (40) are a factor of two with the observed concentration.

Table 2. Observed and predicted concentrations for run 9 experiments.

Test	Downwind distance (m)	Vertical distance (m)	Observed conc. (Bq/ m³)	Predicted conc. Equation (40) (Bq/m³)	Predicted conc. Equation (26) (Bq/m³)
1	100	5	0.025	0.032	0.051
2	98	10	0.037	0.033	0.031
3	115	5	0.091	0.090	0.070
4	135	5	0.197	0.148	0.160
5	99	2	0.272	0.155	0.234
6	184	11	0.188	0.162	0.138
7	165	12	0.447	0.032	0.339
8	134	7.5	0.123	0.033	0.107
9	96	5.0	0.032	0.032	0.034

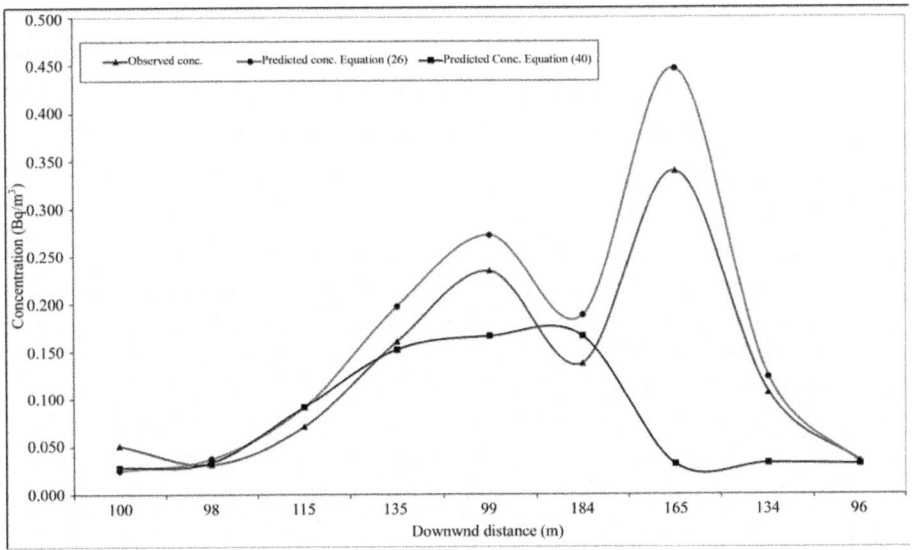

Figure 1. Maximum computed concentrations compared with observed maximum value for each test run Equation (26) and Equation (40).

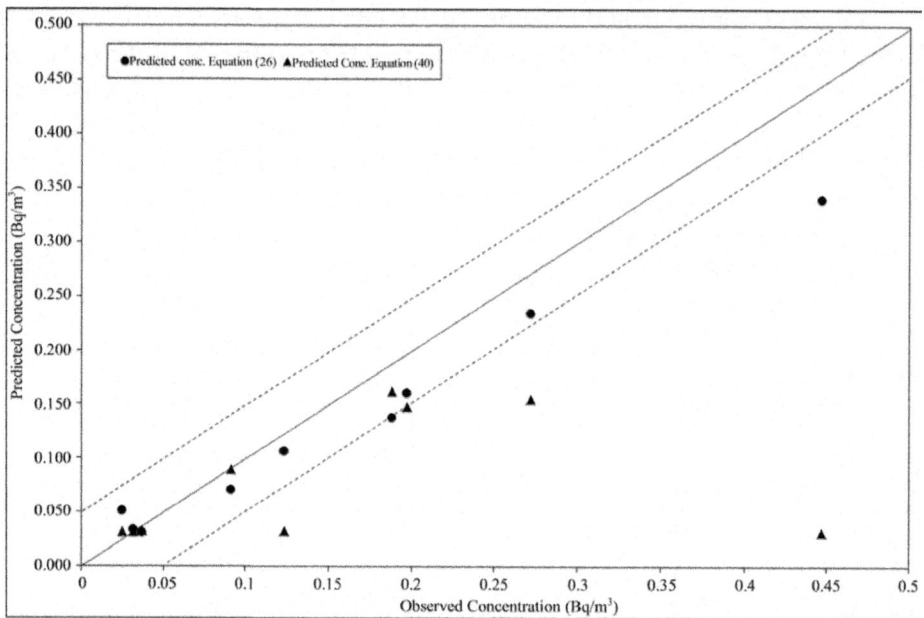

Figure 2. Diagram of predicted model for Equation (26) and Equation (40) with corresponding observation. Solid lines indicate one to one and dashed lines a factor of two.

STATISTICAL METHOD

- Normalized mean square error (NMSE): It is an estimator of the overall deviations between predicted and observed concentrations. Smaller values of NMSE indicate a better model performance. It is defined as:

$$\text{NMSE} = \frac{\overline{\left(C_o - C_p\right)^2}}{\overline{C_o}\,\overline{C_p}}.$$

- Fractional bias (FB): It provides information on the tendency of the model to overestimate or underestimate the observed concentrations. The values of FB lie between −2 and +2 and it has a value of zero for an ideal model. It is expressed as:

$$\text{FB} = \frac{\left(\overline{C_o} - \overline{C_p}\right)}{0.5\left(\overline{C_o} + \overline{C_p}\right)}.$$

- Correlation coefficient (R): It describes the degree of association between predicted and observed concentrations and is given by:

$$R = \frac{\left(C_o - \overline{C_o}\right)\left(C_p - \overline{C_p}\right)}{\sigma_o \sigma_p}.$$

- Fraction within a factor of two (FAC2) is defined as:

  FAC2 = fraction of the data for which

$$0.5 \le \left(C_p / C_o\right) \le 2$$

where σ_p and σ_o are the standard deviations of C_p and C_o respectively. Here the over bars indicate the average over all measurements (N·m). A perfect model would have the following idealized performance: NMSE = FB = 0 and COR = FAC2 = 1.0.

From the statistical method of Table 3, we find that the predicted concentrations Equation (26) and Equation (40) for ^{135}I lies inside factor of 2 with observed data. Regarding to NMSE, FB and COR the predicted concentrations Equation (26) for ^{135}I is better with observed data more than predicted concentrations Equation (40), this is because in model of Equation (26) the wind speed and eddy diffusivity were used as functions in the vertical height z, contrast that Equation (40) the wind speed and eddy diffusivity were used as constant.

Statistical functions	^{135}I			
	NMSE	FB	COR	FAC2
Predicated concentrations Equation (40)	1.75	0.65	0.29	0.74
Predicated concentrations Equation (26)	0.10	0.19	0.99	0.4

CONCLUSIONS

In this paper, we have formulated a mathematical model for dispersion of air pollutants in moderated winds. The diffusion in vertical height direction and advection along the mean wind are taking into account. The eddy diffusivity and the wind speed are assumed to be constant times and variable times. The analytical model is compared with data collected from nine experiments conducted at Inshas, Cairo (Egypt). One gets the predicted concentration Equation (40) that is in poor agreement with the corresponding observation in contrast Equation (26) that gives good agreement with the corresponding observation. Because the eddy diffusivity and the wind speed were used as constants (Equation (40)). On the other hand, the eddy diffusivity and the wind speed had been used as functions in vertical height "z", in Equation (26).

Statistical method also shows that wind speed and eddy diffusivity are taken as a variable better than as a constant.

REFERENCES

1. Wark, K. and Waner, C.F. (1981) Air Pollution. Its Origin and Control. Harper and Row, New York.

2. Wilson, D.J. (1981) Along-Wind Diffusion of Source Transients. Atmospheric Environment, 15, 489-495. http://dx.doi.org/10.1016/0004-6981(81)90179-7

3. Van Ulden, A.P. and Hotslag, A.A.M. (1985) Estimation of Atmospheric Boundary Layer Parameters for Diffusion Applications. Journal of Climate and Applied Meterology, 24, 1196-1207. http://dx.doi.org/10.1175/1520-0450(1985)024<1196:EOABLP>2.0.CO;2

4. Pasquill, F. and Smith, F.B. (1983) Atmospheric Diffusion. 3rd Edition, Wiley, New York.

5. Seinfeld, J.H. (1986) Atmospheric Chemistry and Physics of Air Pollution. Wiley, New York.

6. Sharan, M., Singh, M.P. and Yadav, A.K. (1996) Mathematical Model for Atmospheric Dispersion in Low Winds with Eddy Diffusivities as Linear Functions of Downwind Distance. Atmospheric Environment, 30, 1137-1145. http://dx.doi.org/10.1016/1352-2310(95)00368-1

7. Moreira, D.M., Tirabassi, T. and Carvalho, J.C. (2005) Plume Dispersion Simulation in Low Wind Conditions in the Stable and Convective Boundary Layers. Atmospheric Environment, 30, 3646-3650.http://dx.doi.org/10.1016/j.atmosenv.2005.03.004

8. Cosemans, G., Kretzchmar, J. and Maes, G. (1992) The Belgian Emission Frequency Distribution Model IFDM. Proceedings of the DCAR Workshop on Objectives for Next Generation of Practical Short-Range Atmospheric Dispersion models, Riso, 149-150.

9. Olesen, H.R., Lofstorm, P., Berkowicz, R. and Jensen, A.B. (1992) An Improved Dispersion Model for Regulatory Use: The OML Model. In: van Dop, H. and Kallos, G., Eds., Air Pollution Modeling and Its Application IX, Plenum Press, New York. http://dx.doi.org/10.1007/978-1-4615-3052-7_3

10. Hanna, S.R. and Chang, J.C. (1993) Hybrid Plume Dispersion Model (HPDM) Improvements and Testing at Three Field Sites. Atmospheric Environment, 27A, 1491-1508. http://dx.doi.org/10.1016/0960-1686(93)90135-L

11. Carruthers, D.J., Edmunds, H.A., Ellis, K.L., McHugh, C.A., Davies, B.M. and Thomson, D.J. (1995) The Atmospheric Dispersion Modeling System (ADMS): Comparisons with Data from the Kincaid Experiment. International Journal of Environment and Pollution, 5, 213-228.

12. Gryning, S.E., Holtslag, A.A.M., Irwin, J.S. and Sivertsen, B. (1987) Applied Dispersion Modeling Based on Meteorological Scaling Parameters. Atmospheric Environment, 21, 79-89. http://dx.doi.org/10.1016/0004-6981(87)90273-3

13. Holtslag, A.A.M. and Nieuwstadt, F.T.M. (1986) Scaling the Atmospheric Boundary Layer. Meterology, 36, 201-209. http://dx.doi.org/10.1007/bf00117468

14. Palazzi, E., De Faveri, M., Fumarola, G. and Ferraiolla, G. (1982) Diffusion from a Steady Source of Short Duration. Atmospheric Environment, 16, 2785-2790. http://dx.doi.org/10.1016/0004-6981(82)90029-4

15. Essa, K.S.M., Mina, A.N. and Higazy, M. (2011) Analytical Solution of Diffusion Equation in Two Dimensions Using Two Forms of Eddy Diffusivities. Romanian Journal of Physics, 56, 1228-1240.

16. Essa, K.S.M. and El-Otaify, M.S. (2007) Mathematical Model for Hermitized Atmospheric Dispersion in Low Winds with Eddy Diffusivities Linear Functions Downwind Distance. Meteorology and Atmospheric Physics, 96, 265-275. http://dx.doi.org/10.1007/s00703-006-0208-5

17. Irving, J. and Mullineux, N. (1959) Mathematics in Physics and Engineering. Academic Press, New York.

18. Gradshteyn, I.S. and Ryzhik, I.M. (1965) Table of Integrals, Series and Products. 7th Edition, Academic Press, New York, 1160.

19. Golder, D. (1972) Relation among Stability Parameters in the Surface Layer. Boundary Layer Meteorology, 3, 47-58. http://dx.doi.org/10.1007/BF00769106

20. Essa, K.S.M., Mubarak, F. and Khadra, S.A. (2005) Comparison of Some Sigma Schemes for Estimation of Air Pollutant Dispersion in Moderate and Low Winds. Atmospheric Science Letters, 6, 90-96. http://dx.doi.org/10.1002/asl.94

CITATION

CHAPTER 1

Hsiao-Lan Liu and Yu-Sheng Shen, The impact of green space changes on air pollution and microclimates: A case study of the Taipei metropolitan area, Sustainability 2014, 6(12), 8827-8855; doi:10.3390/su6128827

CHAPTER 2

A.M.O. Abdul Raheem and F.A. Adekola (2011). Air Pollution: A Case Study of Ilorin and Lagos Outdoor Air, Indoor and Outdoor Air Pollution, Prof. JosÃ© Orosa (Ed.), ISBN: 978-953-307-310-1, InTech, DOI: 10.5772/16845.

CHAPTER 3

Xiaopeng Guo, Xiaodan Guo, and Jiahai Yuan (2015). Impact analysis of air pollutant emission policies on thermal coal supply chain enterprises in china, Sustainability 2015, 7(1), 75-95; doi:10.3390/su7010075

CHAPTER 4

David O. Omole and Julius M. Ndambuki, Sustainable living in africa: case of water, sanitation, air pollution and energy, Sustainability 2014, 6(8), 5187-5202; doi:10.3390/su6085187

CHAPTER 5

Dongyong Zhang, Junjuan Liu, and Bingjun Li (2014). Tackling air pollution in china—what do we learn from the great smog of 1950s in London, Sustainability 2014, 6(8), 5322-5338; doi:10.3390/su6085322.

CHAPTER 6

Mölders, N., Butwin, M., Madden, J., Tran, H., Sassen, K. and Kramm, G. (2015) Theoretical Investigations on Mapping Mean Distributions of Particulate Matter, Inert, Reactive, and Secondary Pollutants from Wildfires by Unmanned Air Vehicles (UAVs). Open Journal of Air Pollution, 4, 149-174. doi: 10.4236/ojap.2015.43014.

CHAPTER 7

Cheng, Y., Kang, J., Liu, F., Bassig, B., Leaderer, B., He, G., Holford, T., Tang, N., Wang, J., He, J., Liu, Y., Liu, Y., Liu, J., Chen, X., Gu, H., Ma, X., Zheng, T. and Jin, Y. (2015) Effectiveness of an Indoor Air Pollution (IAP) Intervention on Reducing IAP and Improving Women's Health Status in Rural Areas of Gansu Province, China. Open Journal of Air Pollution, 4, 26-37. doi: 10.4236/ojap.2015.41004.

CHAPTER 8

Al-Hassen, S., Sultan, A., Ateek, A., Al-Saad, H., Mahdi, S. and Alhello, A. (2015) Spatial Analysis on the Concentrations of Air Pollutants in Basra Province (Southern Iraq). Open Journal of Air Pollution, 4, 139-148. doi:10.4236/ojap.2015.43013.

CHAPTER 9

Abdel-Salam, M. (2015) Significance of Personal Exposure Assessment to Air Pollution in the Urban Areas of Egypt.Open Journal of Air Pollution, 4, 1-6. doi: 10.4236/ojap.2015.41001.

CHAPTER 10

Islam, M., Majumder, S., Mamun, A., Khan, M., Rahman, M. and Salam, A. (2015) Trace Metals Concentrations at the Atmosphere Particulate Matters in the Southeast Asian Mega City (Dhaka, Bangladesh). Open Journal of Air Pollution, 4, 86-98. doi: 10.4236/ojap.2015.42009.

CHAPTER 11

Marrouf, A., Essa, K., El-Otaify, M., Mohamed, A. and Ismail, G. (2015) The Influence of Eddy Diffusivity Variation on the Atmospheric Diffusion Equation. Open Journal of Air Pollution, 4, 109-118. doi:10.4236/ojap.2015.43011.

INDEX